A Practical Guide to

RABBIT RANCHING

Raising Rabbits for Meat and Profit

DEBORAH MAYS

A Practical Guide to Rabbit Ranching

CompanionHouse Books™ is an imprint of Fox Chapel Publishing.

Project Team
Managing Editor: Gretchen Bacon
Acquisitions Editor: Shelley Carr
Editor: Joseph Borden
Designer: Mary Ann Kahn
Indexer: Jay Kreider

ISBN 978-1-62008-364-2

Library of Congress Control Number: 2019950100

This book has been published with the intent to provide accurate and authoritative information in regard to the subject matter within. While every precaution has been taken in the preparation of this book, the author and publisher expressly disclaim any responsibility for any errors, omissions, or adverse effects arising from the use or application of the information contained herein. The techniques and suggestions are used at the reader's discretion and are not to be considered a substitute for veterinary care. If you suspect a medical problem, consult your veterinarian.

Fox Chapel Publishing
903 Square Street
Mount Joy, PA 17552

www.facebook.com/companionhousebooks

We are always looking for talented authors. To submit an idea, please send a brief inquiry to acquisitions@foxchapelpublishing.com.

Printed and bound in China
Second Printing

CONTENTS

PREFACE

My college degree is in wildlife biology, and I worked as a wildlife biologist, zookeeper, and veterinary technician, as well as in wildlife rescue and rehabilitation, before finally settling on a career as a research specialist in the biomedical sciences. Though I loved my job, I missed working with animals. So with my husband, John (Beau) Mays, we bought a 100-acre farm that he immediately dubbed Chigger Ridge Ranch. It was head-high in brambles when we moved in; if we could have found a market for chiggers, we would have made a fortune! Since that didn't seem likely, we began experimenting with goats, chickens, ducks, and even mini-cows before deciding that hair sheep were best suited to our land. Thus, we created our "Luscious Lamb" label, selling the packaged meat locally at farmer's markets.

It was one sunny afternoon while at the market that a favorite customer asked rather desperately where he could buy "game meat," as he wasn't a hunter and he was trying to follow his doctor's recommendation to switch to game. Knowing the doctor probably didn't mean the corn-fed, farm-raised (super-expensive) deer available on the Internet, we began to explore what might meet this need. Next thing we knew, we had launched Chigger Ridge Rabbits and began learning about this amazing animal that has the lowest fat and cholesterol of any agricultural animal. Even better, rabbits eat alfalfa-rich pellets and hay rather than corn and grains. This makes rabbit meat even more heart-healthy

with a super omega-6 to omega-3 ratio. Rabbits are notable among farm livestock for their unusual estrus cycle and high birth rate, their unique physiology, their awesome feed efficiency, and the fact that the US government regulates them under different laws depending on whether you are raising them for pets, meat, breeding stock, or research.

Within two years, a farmer can evaluate 60–80 kits from a single doe rabbit, compared to maybe 1 calf from a cow, 3–4 offspring from sheep or goats, or 10–20 piglets from a sow. Even small changes in farm management practices, feed, or genetic selection of breeding stock will result in rapidly seen improvements. "You can't manage what you don't measure" is our motto at Chigger Ridge, and this is reflected throughout this book by charts, graphs, and tables to make it easier to see not just the how but the why of our suggested practices.

We raised rabbits on a commercial scale for many years as Chigger Ridge Rabbit Ranch. We marketed them through commercial rabbit meat processors, restaurants, farmer's markets, local groceries, the pet food industry, and as breeding stock. We had more market than rabbits available—constantly. (Those who claim rabbit ranching for meat cannot be a viable enterprise until more demand is out there are just plain wrong!)

I have a background not only as a veterinary technician and zookeeper working with hundreds of different animal species but also took college courses in animal nutrition, animal pathology, animal physiology, animal parasitology, lab-animal science, and animal production. There was no book of this type available when we started, and much effort was wasted sorting through contradictory information from breeders and self-proclaimed experts. We made our share of mistakes but learned from them how to improve our stock and our bottom line. We hope that this book will assist others to prosper and avoid problems that other animal industries have encountered.

When we first started Chigger Ridge Ranch, we decided hair sheep were the best animals for our land. We later found meat rabbits were an even more viable income source for our operation.

HOW IS THIS BOOK DIFFERENT?

What makes this book different are the words "practical guide." All too often, rabbit-raising books seem to consist of pages and pages of pretty photos of all the different rabbit breeds available, some advice on how to breed them, general plans for hutches, and a bunch of rabbit recipes. On the other end of the spectrum, you might encounter the rare rabbit "textbook," which often has confusing and complex information on rarely encountered diseases and in-depth physiological data but few recommendations on how they apply to the rearing of healthy animals. Such texts give advice on optimal nutrition requirements that would require a Ph.D. in nutrition and your own feed mill to implement.

This book is intended to help people raise rabbits efficiently and humanely as a food source for themselves or others. It is directed to the individual producing safe meat for their family alone, the small farmer trying to diversify, or the large-scale commercial rabbit rancher. Though the book is specific to the United States, as far as wading through laws and regulations, the majority of the book can be applied anywhere in the world.

As a practical guide to raising rabbits for meat or money, this book covers a range of topics, such as the breeds best suited to meat production; how to find a quality breeder; common avoidable ailments; breeding strategies; what to expect in pregnancy, birthing, lactating, and weaning; feeding during different stages of life; various housing options and watering systems; scientific selection of replacement stock; and carcass quality. We incorporate very specific relevant details, such as recommended cage sizes, the pros and cons of various nest boxes, what gauge wire you need for housing and where to find it, how to evaluate a feed tag and supplement your feed when needed, how to protect against predators, and how to keep your barn water lines from freezing.

> **Those who claim rabbit ranching for meat cannot be a viable enterprise . . . are just plain wrong!**

In addition to realistic husbandry advice, this guide illuminates the challenges of running a small farm business, such as how to navigate regulations, taxes, and insurance; how to maximize profits by varying your business plan; how much you should charge for your product; and how and where to market. We spend a whole chapter on all the various marketing outlets available for rabbits and how to approach them. And we give you the tools to evaluate your costs and income and be able to set a product price to keep you in the black.

If you follow the guidance provided in this book, the answer to the question we are asked over and over ("Can I make money raising rabbits?") will most definitely be YES! The answer to the question of "*Will* I make money raising rabbits?" is dependent on many factors, such as how well you research the field, how diligently you pay attention to the needs and health of your stock, how much effort you put into marketing your product, how close you are located to potential buyers and rabbit processors, and most importantly, how much time you have to devote to the enterprise. Rabbits don't raise and market themselves. It can be a time-intensive operation (depending on the size of your herd), but if you put in the effort to keep your animals healthy and happy, are located in a state that is amenable to rabbit agriculture, and are near a potential market, it is hard to *not* make money on such an efficient and prolific creature.

I would like to acknowledge my Content Editors at Fox Chapel, Anthony Regolino, Kerry Bogert, and Joseph Borden, and my Acquisition Editor, Bud Sperry—without whom this book would never have been published. I am also indebted to Dr. Ester van Praag for kindly providing some of the medical photographs included in this work.

Dedication

This book is dedicated to my wonderful husband, John Mays, co-owner and co-manager of Chigger Ridge. Our goal was to produce a "practical guide" for both current and future rabbit ranchers. It is our hope that this work may improve the industry and provide others with a framework for a successful alternative farm enterprise with a real and predictable income.

John Mays with rabbit raised on Chigger Ridge Ranch

CHAPTER 1

MEAT RABBITS AND THE RABBIT INDUSTRY

RABBIT—THE HEALTHIEST MEAT YOU CAN EAT

Rabbit meat is a tender, mild-flavored, all-white meat that can be prepared any way that chicken can—baked, grilled, roasted, sautéed, fried, or stewed. It is low-fat, high protein, very filling, and takes up flavors and marinades well. Commercial rabbit feed generally does not contain added hormones, animal by-products, or antibiotics, and it is more plant-based (alfalfa) than grain-based. This makes rabbit meat higher in omega-3 fatty acids than many livestock species raised today. So if a healthy diet is a consideration in farm production, rabbit actually tops the list (see chart on page 10).

RABBITS *ARE* SUSTAINABLE AGRICULTURE

In addition to the heart-healthy benefits of rabbit meat, the rabbit is a superb example of sustainable agriculture. If managed correctly, it could conceivably feed the planet. As shown in the chart on page 11, rabbits are more efficient at turning feed into edible protein than any animal except the chicken, which it matches.

The rabbit has a feed conversion ratio of 2–4:1, with a mean feed conversion of 3:1. This means that it takes between two to four pounds of feed to convert into one pound of rabbit meat, with the average being three pounds of feed per pound of rabbit. The reason for the variation in feed conversion has to do with differences in breeds, feeds, time of year, etc. (All of these factors affect how efficiently any species of animal converts feed into muscle meat.) Just as an Angus cow is different from a Longhorn cow, different rabbit breeds have different feed conversion ratios. All feeds are not the same in quality and bioavailability. And, of course, in winter more feed is required for an animal to stay warm than in the summer. This chart (page 11) is just for general illustration between species of animals. Careful measurements of feed intake and weight gain on your farm are needed to determine a specific feed conversion ratio for your operation (see Chapter 6). The main source of the rabbit's efficient feed conversion is its unique digestive system, which will be discussed in detail in Chapter 4.

In addition to efficient feed conversion, rabbits are famous for their reproductive capabilities. One doe rabbit is easily able to produce as many as 6 litters of 8 kits or more per year—that is 48+ kits. If the kits are harvested at approximately 4.75–5.75 pounds live weight (the range recommended by many rabbit meat packers), the doe can produce 228–276 pounds of rabbit per year. Not bad, when you consider how little she costs to purchase, house, and feed compared to other livestock.

Comparison of Animal Meat			
SPECIES	**CALORIES PER POUND**	**% PROTEIN**	**% FAT**
Rabbit	795	20.8	10.2
Chicken	810	20.0	11.0
Veal	840	19.1	12.0
Turkey	1,190	20.1	20.0
Lamb	1,420	15.7	27.7
Beef	1,440	16.3	28.0
Pork	2,050	11.9	45.0

From USDA circular #549

Rabbit is lower in calories and fat and higher in protein than any other typically farmed meat.

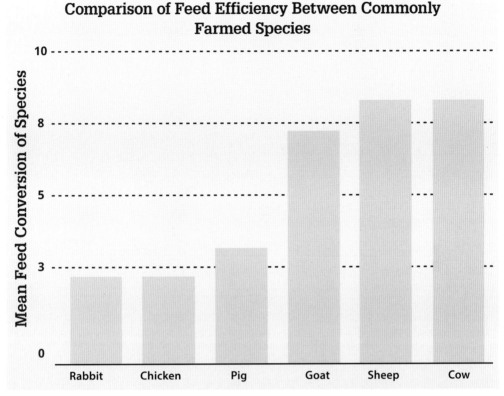

Comparison of Feed Efficiency Between Commonly Farmed Species

Mean Feed Conversion of Species

Rabbit | Chicken | Pig | Goat | Sheep | Cow

Rabbits are three times more efficient than cows or sheep at converting plant-based feed into meat.

Genetic selection of your herd, improvements in feeding strategies, and attention to the health and well-being of your animals may permit an extra litter per year, rearing of 9–10 kits per litter instead of 8, or reaching a higher harvest weight in the same amount of time and feed. These strategies can all increase the pounds of meat per doe rabbit to between 300 and 350. Any way you calculate it, rabbits are an economical way for families in many parts of the world to supplement their pantry, even if agriculture is not their primary source of income. Startup costs are low, compared to most other animal-based agricultural enterprises, and return on investment is *much* faster.

Rabbit farming is also one of the few agricultural enterprises not affected by adverse weather conditions that are a major problem with field crops and grazing animals. When housed correctly, rabbits are not subject to predator animals. They can be raised as a year-round product rather than a seasonal crop and can be easily bred to meet market demands. Thus, they can actually produce predictable monthly agricultural income—a rare and valued commodity in today's farming world. In short, the rabbit is one of the more *controllable* agriculture enterprises.

If managed correctly, rabbits could conceivably feed the planet.

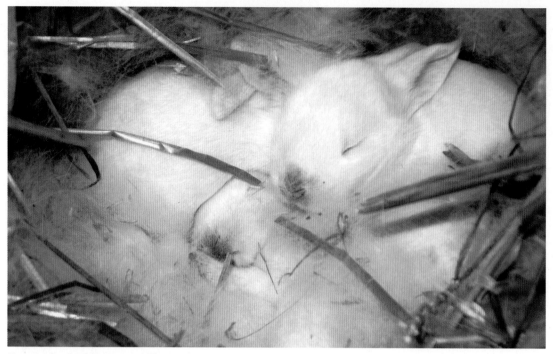
New Zealand Whites are prolific and have fast-growing kits (photo at approximately one week old).

OTHER REASONS TO RAISE RABBITS

Rabbits are quiet, safe, and relatively clean, making them one of the few animals that can be raised on small-acreage farms or even in suburbs or cities, by kids, grandparents, men, or women. These furry models of small-scale agriculture are becoming popular for teaching the next generation about the cycle of life and where food comes from without the need for parents to purchase large tracts of land. With the short time span from birth to harvest, rabbits make awesome 4-H projects and homeschool science and genetics lessons. They have even become part of community garden projects in depressed neighborhoods. Others today opt to raise them to have personal management over a healthy or emergency food supply.

Other Commercial Opportunities Include:

- Worm production
- Soil building directly from the manure
- Selling compost

In addition to the rabbit itself as a product, its manure is highly prized by organic farmers and those wanting to enrich their garden soils, as it does not have to be composted. It won't burn plants if applied directly to gardens. Thus, the rabbit demonstrates sustainable agriculture even more. One working rabbit doe may provide as much as 12 cubic feet, or 300–400 pounds, of manure per year, and that is not even considering manure from her kits. It is drier than cow

Rabbit manure is highly prized by organic farmers, as it is high in potassium and nitrogen.

manure and thus has more nutrients per pound. The composition of the manure varies with the rabbit feed, but since alfalfa is usually a high percentage of any commercial rabbit pelleted feed, rabbit manure is generally high in valuable potassium as well as nitrogen.

The photos below show four tomato plants grown in straight rabbit manure with no additional chemical fertilizer. They produced pounds and pounds of tomatoes for months! Note: never sell your rabbit manure labeled as "fertilizer" unless you are prepared to conduct laboratory tests and component-percentage labeling. Instead, call it organic soil enhancer, soil enricher, or soil dressing. Or you can simply label it rabbit manure.

Worms can also be easily raised under rabbit cages as a sideline product (for fishing bait or for gardens). They turn the rabbit manure into a wonderful rich, black, organic soil, which can also be sold at a premium.

Sideline products from rabbit rearing can pay for a good deal of their upkeep. These tomato plants grown in rabbit manure produced many pounds of tomatoes!

"Every Garden a Munition Plant"

With their efficiency and ease of cultivation, meat rabbits played a vital role in the "victory gardens" of World War II.

Worm and manure sales may be lucrative enough to pay for a good portion of your feed bills in certain seasons of the year. Saving old feed bags to sell this enhanced soil in saves money and is appreciated by your sustainable agriculture customers. (The worms are usually filtered out and sold separately.)

PROBLEMS OF AN INFANT INDUSTRY

So, if rabbits are all that great, why aren't they a major part of the agricultural landscape? Raising domestic rabbits is actually nothing new. They have been raised since 1000 B.C., and since at least 1840 in the United States. During the Great Depression in the United States, they were often the salvation of families with little or no income. During World War II, they were part of the "victory gardens" in Europe and also encouraged by the US government to alleviate the meat shortage. In fact, this very history may be one reason why the rabbit is only just now becoming popular again.

In the past, they were sometimes considered the food of the poor, not the exotic delicacy of today. This "poor man's meat" stigma, coupled with the arrival of television and cute rabbit cartoons in the 1950s, made the idea of putting rabbit on the table less palatable. Note: when marketing your meat rabbit, *NEVER EVER* refer to them as "bunnies"! *NEVER EVER* put pictures of fuzzy baby "wabbits" with big dark eyes on a meat brochure or business card!

Other reasons the rabbit is not yet a major part of today's farming industry is that rabbit meat processors (or live animal buyers) in the United States are not as readily available as with other farm animals. Alternative markets must be researched for your particular area if you are planning on a commercial rabbitry (see Chapters 8 and 9). Farmers must often develop their own market and educate folks on the delicious taste and health benefits of rabbit meat. One cannot depend on being able to dump them at a local sale barn, as is possible with sheep, goats, or cattle. The regulations on packaging and selling rabbit meat also vary by state in the United States and are "gray" and changing. Rabbits are not considered livestock by the federal government and are thus not regulated under the United States Department of Agriculture (USDA); the Food and Drug Administration (FDA) has jurisdiction instead. With its other obligations, the FDA does not really have the time, funding, or expertise to meet this demand, which often leaves regulation and inspection completely up to the states. We will address US rabbit meat regulations and how to cope with them in detail in Chapter 8.

Even with the industry problems, rabbit production is beginning to take off with a growing awareness of the worldwide necessity of efficient utilization of natural resources and sustainable agriculture. The industry is growing rapidly as people look for a renewable nutritious protein source, equal to the chicken but able to be raised in many places where chickens are prohibited. Rabbit agriculture is an answer to the increasing consumer concern for products with low fat, high protein, and no added antibiotics or hormones.

As it matures, the industry must develop animal-rearing protocols that are humane and hygienic, procedures that take into account the social and psychological as well as physical needs of the species. Even then, some people will always argue that rabbits should be in the category of horses and dogs as pets, not food. Indeed, we agree that some individual rabbits have more personality than, say, most sheep or chickens, but are not as engaging or sociable as a lot of goats, cows, or pigs we've known. Rabbits were created and evolved as a prey species at the bottom of the food chain. Their entire psyche is to eat, flee danger, and try to pass on their genes before they are eaten. In the wild, they rarely live over a year before they are dinner for some predator, or before they meet a car on a dark night. In a properly run rabbit ranch, the animals are given food, clean water, and a safe, stress-free environment. They are allowed to breed and raise their young in safety. If animals are not kept happy and healthy, it is not only wrong to raise them for food, they will not produce the nutritious and sustainable protein our own species needs to live.

CHAPTER 2

GETTING STARTED AND RABBIT BASICS

MEAT RABBIT BREEDS AND RABBIT BREEDER SELECTION

Every rabbit guide seems to start out with addressing what breed of rabbit you might want to acquire. As this book is specifically concerned with raising rabbits for meat, we will focus only on what makes a top-quality meat rabbit. *Any* rabbit breed can be used for meat. A "meat breed," however, is considered to be the larger, more muscular breeds—for obvious reasons.

There is much more to consider, however, when selecting a breed that will put food on the table (or lots of tables) in a *cost-efficient* manner. To economically and efficiently raise meat rabbits, whether it is for your family or commercially, you need a rabbit that reaches a slaughter weight of 4.75–5.75 pounds in the shortest amount of time possible (2–4 months). If it takes longer than 4 months, you have a "roaster" versus a "fryer" animal (i.e., less tender dinner, same as in the chicken industry).

Rabbits come in all shapes, colors, and sizes. How do you choose which to raise?

The rabbit grows *very* rapidly from birth to two months, then the growth curve begins to flatten out (discussed more in Chapter 4). In the superior meat breeds, this curve doesn't flatten out as fast, and the animal continues to grow rapidly until that magic harvest range. You can still eat the smaller or slower-growing rabbit breeds that may only reach 3.5 or 4 pounds at 3–4 months, but the meat-to-bone ratio is not what you want for an ample family meal, and definitely will not market well to the public. The carcass quality of these rabbits resembles a squirrel more than a plump, meaty domestic rabbit.

On the other end of the spectrum, you do not want to select the giant breeds. These have an adult weight of 15–20+ pounds and take much more food to maintain and more space for living quarters. Plus, a lot of that extra weight is not necessarily meat, but bone. The muscular mid-size breeds are recommended.

Consider your potential breeding rabbit's physique when selecting or buying. They should have a good long, wide back, as the loin is the tenderest and best cut of the rabbit (see Chapter 7 on rabbit carcass and cuts). They should have large, muscular thighs, as that is where most of the meat is located. But don't neglect the front legs—your rabbit needs to be well balanced. Select a rabbit with a wide rib cage that allows more growth room for kits during pregnancy. The optimal ear size can vary depending on where you live. We prefer larger ears in the hot south, as the heat exchange for a rabbit is mostly in their wide ear surface. We also choose rabbits with less dense coats for the same reason. White-colored rabbits with lighter meat are desired by commercial meat packers, so colored rabbits should be avoided.

Other than fast growth and conformation, a major consideration in deciding on a rabbit breed is how prolific or productive they are. We need to mention the term "maternal breed" here. This refers to the doe or female rabbit's breed. Maternal breeds have been selected for maternal traits. This means characteristics such as how easily they will breed or rebreed after having kits, how many kits in a litter they have, how long their reproductive life is, how consistent they are at making adequate nests and caring for their young, and how much milk they produce. A good maternal meat rabbit breed conceives readily, makes a good nest even with their first litter, has eight or more kits for several years, and raises them all to harvest reliably.

A long, wide back and muscular thighs will result in tender loins and meatier drumsticks.

There are over 50 breeds of rabbit in the United States, from minis to giants. There will always be some controversy over the "perfect" meat rabbit breed. Much of this debate is based simply on anecdotal experience. If you have a really great Californian rabbit breeder down the road, you may be able to get high-quality, well-selected, productive breeding stock from them, whereas others (based on their own experiences) may swear all Californians are small and unproductive.

Regardless, the New Zealand White and Californian rabbit breeds are undeniably the two most common and intensively bred rabbits for meat in the United States. They are large but not considered giants. New Zealands generally run 9–12 pounds adult weight, and Californians are slightly smaller at 8–10 pounds adult weight. Since they are the two most common breeds, it is far easier to find a good breeder of one or the other of these breeds rather than the less prevalent breeds.

By "good breeder," we mean a good *commercial* breeder—not a show breeder. Nothing against rabbit shows, but shows do not judge for maternal traits such as large litters, easy breeders, fast breed-back times, good nesting instincts, or heavy milking ability. Show raisers are not necessarily looking at feed efficiency, strong immune systems, or rapid growth in kits. Yet all of this is absolutely essential for the successful meat rabbit raiser. Far more important than the breed itself is the *breeder* that you select. I would much rather a breeder be able to tell me how fast his kits grow than how many shows he has won.

There are other breeds that can match the New Zealand and the Californian, such as the Champagne D'argent (9–12 pounds adult weight), known for its excellent meat-to-bone ratio and fast early growth, but they are colored and with slightly darker meat that is not desired by the meatpacking plants. Additionally, the Champagne D'argent and most other such breeds are simply not raised in the United States on a commercial scale for meat, and thus it is much harder to find good selected breeding stock to start you off right. In our opinion, why reinvent the wheel? We recommend that you go with the two breeds that have already been selected for nearly a hundred years in the United States as productive commercial meat animals to use for your maternal lines, the New Zealand White or the Californian.

The muscular, mid-size breeds are recommended.

New Zealand Whites are among the most popular meat breeds due to their excellent maternal traits.

Californian with excellent conformation. Note that this breed's fur is white on the meat parts, producing the lighter-colored meat preferred by packing plants.

NEW ZEALAND WHITE RABBITS

The New Zealand White, despite its name, was actually developed in the United States. It was recognized as a breed in 1920. It is a mix of Angora, American White, and Flemish Giant. It is a solid white albino rabbit. A white rabbit has always been preferred by meat packers because it has lighter meat, and some say it is easier to skin and process. This is one reason the New Zealand and Californian became popular meat breeds.

Along with New Zealand Whites, there are also New Zealand Reds, Blacks, Broken coloration, and Blues. These color variations have not had the intensive commercial selection pressures of the New Zealand Whites, however, and it is recommended that the Whites be your choice if you are looking for an efficient operation. Whenever we mention New Zealand throughout this book, we are referring to the New Zealand White. The New Zealand reaches sexual maturity and can be bred between 4–6 months old. They have strong nesting instincts and are excellent milkers, easily able to raise 10 or more kits. They consistently produce eight or more kits per litter and can be productive for up to five years.

CALIFORNIAN RABBITS

The Californian was also developed in the United States in the 1920s and is a mix of the Himalayan and Standard Chinchilla Rabbit with some New Zealand blood thrown in. Like the New Zealand White, they have been extensively selected for maternal traits.

They are white where it counts (i.e., the meat parts), with black only on the "points" (that is, the ears, feet, tail, and nose). People lucky enough to have access to both good New Zealand and Californian commercial breeders to obtain their initial breeding stock may elect to buy both breeds and tap into some of the hybrid vigor that crossbreeding can give (i.e., using a California buck with a New Zealand White doe and vice versa; see Chapter 6 on crossbreeding). Or you might want to keep both breeds purebred and see which does better for you. This will probably be the result of the *rabbit breeder* you buy from and their selection over the years, not necessarily the breed.

HOW TO TELL IF A BREEDER IS GOOD OR NOT?

First and foremost, look for signs of disease in the rabbitry you are considering buying from. Check for signs of:

> **If the breeder won't let you tour the facility, run!**

- Ear mites (dark wax in ears or missing fur around ears)
- Sore hock (sores on bottoms of back feet)
- Weepy eyes
- Sneezing
- Matted hair on the insides of the front legs (caused by the rabbit cleaning a runny nose and indicative of respiratory infections)
- Missing fur or excessive scratching
- Any signs of diarrhea
- Overgrown nails or teeth

Red or dark urine is normal in a rabbit and is not a problem. If the breeder won't let you tour the facility, run! A few may try to tell you that they have a "closed rabbitry" to prevent disease introduction. To that we say they can provide prospective buyers with disposable smocks, shoe covers, and gloves for less than two dollars if they are that worried.

To prevent introduction of disease, avoid breeders who use wooden cages and open water bowls.

Next, look for housing that is sanitary and takes into consideration the animal's comfort. Make sure the housing is well-ventilated, has no smell, has little evidence of rodents or flies, and has a means to keep the animals cool in summer. Look for water systems that are designed to prevent disease transmission—NOT water bowls, which are notorious bacteria breeding grounds, but automatic water systems or at least water bottles. Likewise, feeders should be situated to prevent contamination of feed. Try to find a facility that does not use wood in the hutches or for nesting or resting, as this is notoriously impossible to disinfect well.

After you have decided that the facility is clean and disease-free, you should get some idea of how the breeders maintain productivity.

- Do they keep breeding records to prevent too much inbreeding?
- Do they know at what age their kits hit market weight?
- Can they tell you their average litter size and kit survival rate?
- Are they able to explain the criteria they use in selecting which breeders to keep (or sell to you)?

If they don't keep records and can't answer these basic questions, they are hobby breeders, not serious commercial breeders. Without good productive genes to start off, it is difficult to select for all the traits you need to make your operation cost-effective.

SHOULD YOU PAY MORE FOR A PEDIGREE?

People often ask if they should buy rabbits that come with a written pedigree or are registered as purebreds with various breeder associations, even if they cost much more. To decide this, you should know that a pedigree can be anything a breeder wants to write out. There are plenty of templates available on the Internet for fancy-looking pedigree papers, and there is no way to tell if what is written on the paper is actually the parents of the rabbit you are buying. Even if the breeder is totally honest and the pedigree papers are perfect, what do some rabbit's ancestors' numbers actually tell you? Not much.

However, purchasing a rabbit registered with the American Rabbit Breeder's Association is different from just buying a rabbit with a pedigree. Registration costs $6 per rabbit (at present), and the rabbits must be examined by a registrar when they are at least six months of age. Since we suggest you buy your rabbits before 6 months for optimal breeding in a commercial setting, this requirement may delay breeding and impact productivity. ARBA registration will show that the rabbits have met the minimum breed standards for conformation (which is good), but it tells you nothing about the line's productivity. There is nothing wrong with buying rabbits that have pedigrees provided by the breeder or are ARBA-registered, but in our opinion, it is not really all that helpful in a commercial enterprise.

Once you have found a reputable commercial breeder from whom to buy your initial stock, you may ask, "Why can't I simply buy both new females and males whenever the animals start to lose their productivity instead of raising replacement animals and doing all the selection stuff myself?" For your bucks, this is an option if you have found a breeder that you like within a reasonable distance. The buck can be kept productive for about six years. The doe rabbit is a different story. It is absolutely essential that you be able to cull unproductive females. Some individuals may be productive for 4–5 years, but many start to become less productive by three years. By productive, we mean breeding easily, making a good nest for her 7–10 kits, milking well enough to raise all kits to a good weaning weight, and regaining enough weight herself after rearing her kits to breed back in a reasonable amount of time (we will go into breeding schedules in Chapter 3).

If you necd to buy new does often (at $30–$40 per doe), this can obviously become cost-prohibitive. You can raise your own doe rabbits to breeding age for well under $10. You are also more apt to cull an unproductive doe when required if you do not have to make a trip to buy another but can simply pull one out of your own replacement pens. Every time you buy a new rabbit, you also risk introducing disease to your rabbitry. (Besides all that, half the fun of raising rabbits is watching the improvements in your stock in as little as one year!)

SELECTING YOUR BUCK'S BREED

Once you have decided on your maternal breed, you should think about what breeds of bucks you might like to use. You can stay purebred and use New Zealand or Californian bucks on the same breed of does. You can crossbreed New Zealands and Californians. Or you may want to consider what is known as "terminal sires," where the breeding is considered "terminal"—all the offspring will be slaughtered and used for meat, *not* used for replacement breeders in your herd. Flemish Giant and Altex are examples of two breeds that may be considered as terminal sire rabbits to be bred to your maternal-line (New Zealand or Californian) females.

Why not use these resulting crossbred kits as replacement breeders? Because the Flemish and Altex breeds have not had that extensive selection for maternal production traits that the New Zealand or Californian have had. So a kit descended from these breeds may or may not be a good maternal animal. The terminal sire is *just* to give size and fast early growth to kits *intended for meat* and to provide hybrid vigor, resulting in a robust and hardy meat kit (see Chapter 6 for rabbit breeding genetics). If you use terminal sire breed bucks to raise meat animals, you will still need New Zealand White and/or Californian bucks to produce your replacement females and maintain your purebred maternal breeding stock lines.

Flemish Giants

Flemish Giants are definitely that—giants. They were developed in Belgium in the 16th century and can range from 13–20 pounds as adults. (Due to the heavy bone structure, however, not all of that is added meat.) A whole herd of female Flemish Giants in your rabbitry would eat you out of house and home, but a Flemish buck or two, bred to your New Zealand or Californian does, will give you a very nice, fast-growing, crossbred meat kit. Flemish Giants come in a variety of colors. Since white is preferred for the meatpacking industry, that color is the best choice.

Flemish Giant terminal sires can add weight to your meat kits.

The Altex rabbit was specifically developed to breed kits that showed rapid and sustained early growth.

Altex

The Altex is another breed that was specifically developed as a terminal sire. It was developed by Dr. Lukefahr and his students (Medellin and Lukefahr 2001) in the rabbit research program at Alabama A&M and Texas A&M Universities (thus the name Altex). The Altex is a mix of Californian, Champagne d'Argent, and Flemish Giant. It is colored like a Californian, with white body and darker points (ears, nose, tail, and paws). The Altex rabbit's sole purpose and development was for very rapid and sustained early growth of kits. The measurement used exclusively for selection in the development of the breed was heavy, 70-day kit market weights. This means that Altex doe rabbits were never selected for maternal traits. Indeed, we found the Altex doe harder to breed in general and with a shorter productive life than the New Zealand White. It was also rare that the Altex does we had produced eight or more kits.

If you are a large enough operation, you may want to keep a small herd of a purebred terminal sire breed such as the Altex or Flemish Giant—both females and males, to breed your own replacement terminal sires. It should be noted that this starts adding extra cages and extra work to maintain two or three different breeds.

Altex cross kits get large fast. Note the size of this kit (right) compared to the full-grown New Zealand doe (left).

Many people ask, "Do I *need* to have terminal sires?" The answer is absolutely not. These are used mostly by the larger breeding farms to give a slight increase in finished weight or finish at a target weight a week or 10 days early. When you are talking hundreds or even thousands of kits per month, a week of feed cost savings can be absolutely critical. However, if you are just raising for your family and friends, the added costs of keeping terminal sires available may not be worth it. Instead, you can select your herd bucks from your most productive and fastest-growing maternal breed females and stay with all one breed.

Below is a chart from an experiment conducted in our rabbitry, where we took pure New Zealand litters and crossbred New Zealand by Altex litters and split them up so that good New Zealand doe females were raising four of the purebreds and four of the crossbred kits (thus the milking ability of the doe rabbit didn't affect the results). We weighed each kit of the eight litters in the study (64 kits total) at 3, 20, 30, 50, and 75 days. As you can see in the chart below, there is a slight growth advantage (approximately ½ pound) to the crossbred (New Zealand by Altex) versus the purebred (New Zealand by New Zealand). That is, the purebred New Zealand kit may weigh 5 pounds when the crossbred kit weighs 5.5 pounds. Or put another way, the crossbred kits will hit the minimum market weight of 4.75 pounds about 7–10 days earlier than the purebred kits. If you are selling commercially and your buyer wants a *minimum* 5.5-pound rabbit at harvest, and since this is the time when the animal is eating the most and growing the slowest, it is obvious that crossbreds using terminal sires are cost-effective.

Even though the crossbreds are a commercial advantage, one mistake a lot of people starting out make is trying to handle too many breeds at once. If you want to raise your own terminal sires (say Altex) and not have them too inbred, you will need a *minimum* of four Altex does and two bucks to breed with them. Since Altex females are not as prolific in general as New Zealand or Californians, this means you will have at least four cages of less productive females. By less productive, we mean that these females may not conceive as often,

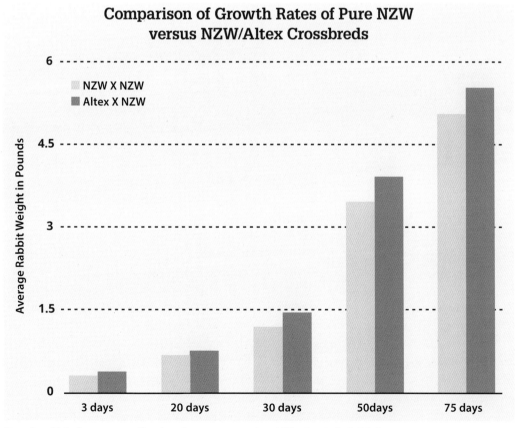

Crossbred kits have a growth advantage over purebred kits due to hybrid vigor.

reject litters more often, fail to make nests, or have only 4–5 kits instead of 8. You have to consider whether or not the use of terminal sires and the hybrid vigor and faster growth they give your meat kits is going to *pay* for a number of cages being less productive, just to maintain a separate breed line to draw on for an occasional terminal sire. The terminal breeds are also larger and require more feed to maintain breeders.

The best solution is to try to find a specialty breeder that has Altex, Flemish Giants, or another suitable terminal sire breed near enough to you that you can purchase a new terminal sire buck every few years and keep most of your cage space for your productive maternal breed rabbits. This is especially true if you are a small family farm operation where every cage needs to be as productive as possible.

THE BEST AGE TO BUY BREEDING STOCK

Some people believe they should purchase only "experienced" senior does that have already had kits, proving she has performed in the past. Others want to buy newly weaned rabbits at a month old so that they can watch them grow up and the animal will be totally comfortable with their surroundings when they have kits. Below is our opinion on the optimal age to buy a rabbit and why. This opinion will vary with any breeder you talk with, but they should be able to give a good reason for the age they choose to sell you.

We sold thousands of breeding stock rabbits and, except for very special circumstances, always sold them between three and five months of age. A young rabbit less than three months is susceptible to gastrointestinal tract disorders. This is the age when antibodies the kit received from the mother are being lost and the kits are still developing their own immune system. They are also susceptible to stress from weaning to three months, as they are learning their environment without the security of the nest. They are moving to solid foods as well, which is a systemic stress. Their "good" intestinal bacteria loads are still developing and stabilizing at under three months. Rabbits that do develop gastrointestinal problems younger than three months and the associated diarrhea that accompany them are susceptible to death. For these reasons, we never sold, and don't recommend that anyone buy, a rabbit younger than three months. The best age breeding stock to purchase is between three and five months old.

The five-month upper limit is for the New Zealand White and Californian. (It can be a bit longer for the giant breed rabbits.) The reason for this upper limit is not health related but rather breeding related. You should quarantine your rabbit for three weeks before breeding them. That will make a newly

> **The best age breeding stock to purchase is between three and five months old.**

purchased 5-month-old rabbit nearly 6 months old before being bred for the first time. Beyond six months, the rabbit may become too obese if they are not in an active breeding program. Also, if you purchase a rabbit that is 4–5 months old rather than one that's 2–3 months old, you are less likely to get one that an unscrupulous breeder simply pulled out of his meat animal pens that he had planned to sell for meat, but rather one that has been saved out specifically as a replacement breeder for himself or to sell as a breeder. A meat rabbit at 4–5 months old should be way too heavy at that age for the normal fryer meat market.

Why not purchase an older, experienced doe that has already had a litter or two? In some cases, you may get a good buy on this. The breeder may have a good doe that he would not normally cull, but he has too many of the same bloodline and wants to try other genetic mixes. However, you should know that a commercial rabbit raiser already has time and money invested in an adult rabbit. It has been raised to breeding age and beyond, and the risk taken that it will breed, nest, and nurse. You will usually be charged extra for a prime breeder, or you may get a cull breeder whose performance is really not up to standards.

HANDLING AND SEXING

Despite seeing magicians pulling white rabbits out of their hats by the ears when we were kids, you *do not ever* pick up a rabbit by its ears. Just because they don't struggle much when doing this doesn't mean it isn't painful. There is no animal on the planet that likes to have its full weight suspended by its ears!

"Scruffing" of the rabbit is our preferred method of handling if done correctly. This is when the loose skin on the back of the neck is used to *initially* lift the rabbit, and is the same method a female dog or cat will use to lift its pups and kittens. A small rabbit less than three pounds can even be carried for a short distance by the scruff, as it is not heavy enough for this to be a problem for it. However, a heavier rabbit will need additional support to the hindquarters immediately after lifting. After lifting a heavier rabbit, you must secure it close to your body and support its hindquarters. Rabbits will not bite you (except perhaps a very protective doe defending her nest), but they will often scratch you with their powerful back legs. As they are prey animals and their instinct is to be "free to flee," few rabbits like to be picked up or carried unless they are pets or show animals and it is a daily occurrence. As seen in the photos below, the rabbit can be tucked under one arm or held firmly against your chest. The object is to make them feel secure but not overly restrained or confined and to keep their paws (and claws) away from your hands and face. It is recommended to always keep one hand at least lightly on the scruff to allow you to maintain control if the animal suddenly struggles.

> You *do not ever* pick up a rabbit by its ears.

A few people may not believe in scruffing for picking up a rabbit and say they should only be picked up by the body like you would a cat or small dog. But unlike these animals, a rabbit can kick out

Proper handling: one hand securing the hindquarters and one hand for control at the scruff of the neck.

Holding a rabbit for sex determination

Sexing adult female

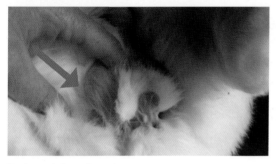

Sexing adult male

Sexing immature male

unexpectedly, and sometimes violently enough to actually injure its back. Handling thousands of rabbits over the years, we never had an injury to the rabbit using our method of handling.

To determine the rabbit's sex, you will need to have the rabbit turned slightly on its back. You will also need a free hand to part the hair to look at the genital region. You can still use the scruffing method to accomplish this, as seen in the above photo panel. You will need a second person to do the sexing, as both hands are required to hold the rabbit. If you do not have a second person, you can place the rabbit's rump on a table, freeing up one hand so that you can part the hair and look for the sex organs.

Finding the sex organs in an adult rabbit is relatively easy. The adult male has visible testicles when the hair is parted (see orange arrow in photo above), and when pressing lightly with your index finger above the penis, it will extend. In the adult female, there are, of course, no testicles, and when pressing lightly with the index finger, you will see a slit rather than a penis emerge.

Sexing the juvenile rabbit before it reaches sexual maturity is a bit more difficult. In meat rabbits that are below 3–4 months of age, in general, the testicles in the male have not yet "dropped" out of the body cavity and cannot be seen. However, when pressing with the index finger above where the penis or vagina should be, you will hopefully see a small "tube" emerge in the male and a small slit in the female. Unfortunately, the immature penis will not always "pooch out" out very far and can be mistaken for a slit of the female fairly easily. Even very experienced rabbit raisers can misidentify the sex in young rabbits, and the younger they are, the more likely this is to occur.

It is recommended that if you sex rabbits before they reach full sexual maturity, you should check them again as they reach four months of age. At this age, many meat rabbit breeds can become sexually active. If you think you have a cage of females and you have inaccurately sexed one, and it is a male, you may have some unexpected pregnancies. If you have accidentally left two males together, there may be severe fighting and injuries as they hit sexual maturity.

QUARANTINE AND INITIAL DAYS WITH A NEW RABBIT

When you bring home your new rabbit, it should be immediately placed in quarantine. Yes, this is essential! Your rabbits will likely be housed in close proximity to each other, and a new animal that is stressed with a move may develop and spread disease. We recommend that the new rabbit be quarantined for a minimum of three weeks. Always feed and handle your quarantined animals last, and then change your clothes and wash your hands so that you will not accidentally vector diseases from your quarantined animals to your other stock.

Rabbits are actually more sensitive to changes in their environment than many people realize and should never be brought home and immediately bred or stressed. Move quietly and speak softly around the new animals. Do not allow young children or pets to go near (even if little Johnny is begging to see the new bunnies). One of the saddest things to see is a rabbit with a broken back. Even in a cage that the rabbit is familiar with, it can startle and have an adrenaline rush where it instinctively runs. Unfortunately, in a cage with no place to flee, it bolts around the cage in a circle, bouncing off the sides. If your rabbit starts to do this, freeze all action and remain quiet or speak softly to it. In this adrenaline-charged state, it can run headlong into a side of the cage and break its neck or back. A new rabbit is more prone to do this than a rabbit that is familiar with its surroundings. Always announce your presence to your rabbits before entering the rabbitry, and have strict rules for children around the rabbits—no running, loud voices, hitting the sides of the cage, or carrying strange objects.

Identifying rabbits with a broken back is easy. They usually break it mid-spine and will drag their hindquarters. They often do not appear to be in any pain. If there is some movement to the hind feet, there is a possibility the back is just injured and not broken, in which case the rabbit may recover (you will see more and more movement over the next 24 hours). But in 95 percent of the cases, it is permanent paralysis, and the animal must be euthanized. If a rabbit breaks its neck during such an episode, the result is invariably instant death.

While in quarantine, your new rabbit's feed and water intake needs to be monitored carefully. A change in feed may be refused at first. You should always obtain some feed from the rabbit breeder and switch your rabbit over slowly from the old feed to its new pellets, mixing them in a bit more each day. Always provide hay to a new rabbit to prevent gastrointestinal (GI) stasis with the stress of the move. It should be noted that with the very low fat reserves of a rabbit, just a few days without food can cause severe weight loss and endangered health. Think of just one day without food for a rabbit as being equivalent to you not eating for 3–4

A veterinarian may be needed to help you diagnose diseases correctly—especially when you are first starting out.

days. A rabbit will not eat if it cannot drink, so though it sounds basic, you must make sure your new rabbit understands the watering system in its new home. If it is used to water bowls, it may try a water bottle and get a face-full and be afraid of it. If it is used to water bottles, it may not know how to use the valves of an automatic watering system. If you are unsure, provide a water bowl to be on the safe side (even though, as seen in later chapters, we are strictly against water bowls as a regular water source due to disease transmission).

COMMON AILMENTS AND THEIR MANAGEMENT

After much thought, we decided not to include in this book a separate chapter on all the possible rabbit diseases that may occur in a rabbitry. We believe that this would encourage a rabbit raiser to try to diagnose diseases without adequate knowledge and experience. We urge you instead to develop a relationship with your local veterinarian. Yes, it will often be uneconomical to take every rabbit showing any symptoms of disease to the veterinarian, but it is important to learn to diagnose correctly to forestall the spread of diseases. We will discuss throughout this book ways to *prevent* many common diseases. We cannot stress enough that most diseases in commercial rabbitries are the result of improper management and, instead of being treated, should be prevented! If you do not have a local veterinarian who is familiar with rabbits, there are a few internet sites that provide professional information. Be sure the site is produced by someone with veterinary training. (We all know the trap of following the advice of "experts" who are anything but!) One site that this author considers both practical and completely trustworthy is Dr. Ester van Praag's MediRabbit site (www.medirabbit.com), where there are numerous health-related articles posted.

Hopefully, you have paid attention to the beginning of this chapter and toured the breeding farm before purchasing your rabbits. Even if everything looked wonderful at the breeding facility, the first thing to do when you get your new rabbit home is to give it a thorough check from end to end before even putting it in its quarantine cage. Ask yourself the following questions:

Though not fancy quarters, these kits are being raised in a clean cage with automatic water. There are no wood surfaces that are difficult to disinfect. They are being fed their required hay in a sanitary manner. They show no signs of diarrhea or respiratory disease. They are active, eating, not overcrowded, and have clean, glossy coats. This is what you look for when selecting your breeder.

- Is there waxy material in the ear canal or any signs of irritation?
- Is there any sneezing or discharge from the eyes or nose?
- Is there any staining around the anus, indicating diarrhea?
- Are the teeth aligned and normal looking?
- Are the nails overgrown, or are there any lesions on the bottoms of the feet?
- Is the skin free of lumps, abscesses, scabs, or missing fur?
- Is there any discharge from the genital regions? This is rare, but never normal.

Weigh the rabbit and record its weight so you can tell if it loses or gains weight in the quarantine time period. If any of the preceding symptoms are present or develop during quarantine, you need to think carefully before you even consider introducing the new rabbit to your other animals. The most economically important common problems that result from these symptoms will be discussed in detail in the following pages. There are additional, less common bacteria, parasites, and viruses that can invade your rabbit herd, and again, we encourage veterinary assistance if you are unsure of a diagnosis. There are very few antibiotics that a rabbit can tolerate and no approved vaccines for any health condition in the United States. We cannot stress enough that proper husbandry and prevention are key.

EAR MITES

Ear mites are small, highly contagious parasites that live in a rabbit's ears (contagious to other rabbits, not to humans, as we are not their natural hosts). Even if the ears look perfectly clean, ear mites are so common that we suggest prophylactically treating the rabbit as you transfer it from your transport cage to quarantine, and then once a week before you put it with your other rabbits. Ear mites are a common ailment of most meat rabbitries because there are a large number of rabbits kept close together.

Ear mites are microscopic, and when they bite the rabbit, it causes an allergic reaction and itching. At first, all you will see of an infestation is a bit of brown, waxy material deep down in the ear canal. If left untreated and allowed to multiply, the mites will begin to destroy the ear tissue—and the rabbit's constant scratching does too. Eventually, the ear will become thickened, covered with grayish brown crusty material, inflamed, and possibly infected. Rabbits with severe infestations will go off feed as well (which is always dangerous, but especially so in pregnant and nursing animals). Ear mites are not a minor irritation; they can become a serious health threat. If you suspect ear mites in a rabbit and want to confirm it, use a cotton swab soaked in mineral oil to swab under the bit of crusty material and take it to your veterinarian. Ear mites can live off their rabbit host in the environment for up to 72 hours.

So what do you treat them with? You can ask 10 rabbit raisers this question and probably get 10 different answers. We will tell you what *we* used to control mites and why. First of all, remember you are raising *meat* rabbits—that is, rabbits you and other people will be consuming. Many of the remedies you will hear others discuss are not labeled for rabbits, and it is unknown how much will be absorbed into the rabbit's system or how long it will stay there. Remember that a female rabbit will be either pregnant or nursing for most of her life, and insecticides can be transmitted in the uterus to the fetal rabbits and also through the milk to growing kits. Fortunately, the strong immune protection from the mother to the baby rabbit

Left untreated, ear mite infestation can lead to severe allergic reaction, pain, and infections. This can happen in an astonishingly short period of time.

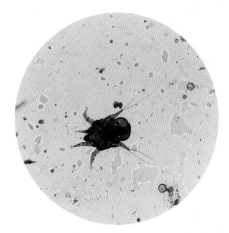

Ear mites may be a microscopic parasite, but they can seriously impact production!

usually protects it from severe allergic reaction for at least the first two or three months of life. You rarely find a severe mite problem in a very young rabbit unless they are in overcrowded, unsanitary, or other stressful conditions. Since harvest for meat is between two and four months of age for meat rabbits, and the mite life-cycle is 21 days, mites are usually not much of a problem for your "grow-out" rabbits that you are finishing for food. This means that your main treatment concern will be your adult doe and buck rabbits.

A popular drug you will see mentioned on the Internet is ivermectin injections, which are reputed to be effective. Ivermectin is not (as of the writing of this book) labeled for rabbits, however, and there is, therefore, no known meat withdrawal time. If you use ivermectin (or any other systemic drug), legally, you must reveal this when taking the rabbits to a processor for slaughter or selling the animal as a meat animal to a buyer. Insecticide companies do not have enough rabbit customers to do the necessary trials to find safe levels for the animal or for the human that will be consuming it. You cannot use other animals' drug withdrawal times for rabbits, as every animal has different drug processing rates.

Our preferred treatment is mineral oil or baby oil placed directly in the ear canal. The oil simply smothers the mite and kills it, so there are no insecticides involved. It will not kill the mite eggs, however. Since the life cycle of the mites is 21 days, you must treat twice a week for at least 3–4 weeks to break the life cycle in an active infestation. This is the safest treatment for a rabbit, even if it is a pain in the butt (we mentioned that rabbit raising can be time consuming, remember). The oil also softens up the thickened ear and allows the crust to fall off in bad infestations. The ear of even severely affected rabbits will look 90 percent better within two days of the first oil treatment. One thing to keep in mind is that when you stick any liquid in an animal's ear, it will shake its head immediately (and thereby shake mites onto surrounding rabbits).

If you want to clear up the mites faster and more reliably, you can use a miticide. We believe that pyrethrin ear mite medicine is the safest to use. This is applied directly in the ear like the mineral oil to kill the mites on contact. It is not a *systemic* insecticide. Pyrethrins have a very low rate of absorption through the skin. It is also not significantly absorbed through the gastrointestinal (GI) tract if the rabbit cleans its ears and ingests the medication. You can purchase this labeled for rabbits (usually 0.06–0.15 percent pyrethrin). We will note that the ones labeled *only* for rabbits are usually for pet rabbits and come in very tiny bottles that will be cost prohibitive for a major rabbitry, so look for those labeled for multiple species, or use mineral oil to mix with a stronger pyrethrin to bring it down to a safe range for rabbits. Since, as mentioned above, pyrethrin is not readily absorbed through the skin or GI tract, it is hard to overdose, even if you accidentally drop in too much. That being said, it has been reported that severe overdosing with pyrethrins can be neurotoxic and result in seizures. We never observed this phenomenon in many years of use.

If you do choose to use an ear mite medication, it is best to use only in dry, non-pregnant does or bucks unless the infestation is severe enough that the animal needs immediate relief that you feel mineral oil alone will not give. This is why we recommend treating during quarantine, as that will be one of the few times your new female rabbit will not be in some stage of pregnancy or nursing.

Ear mites can become severe surprisingly quickly. If ear mites are present at all in your rabbitry, we recommend that every rabbit be examined for them on at least a weekly basis. This means looking *down* into the ear canal—not just checking at eye level from outside the cage.

Though ear mites are not dangerous enough to warrant euthanasia of a new rabbit that shows symptoms, if you are lucky enough to be free of this parasite in your herd, we recommend doing everything possible during quarantine to make sure that you do not introduce them. They can significantly impact your productivity, be painful and irritating to your rabbits, and are extremely time-consuming for a farmer to keep in check. This may mean extended quarantine time for a new rabbit that you suspect harbors them and the ivermectin injections mentioned previously (talk to your veterinarian for dosages or alternative treatments).

SNUFFLES

A second highly contagious disease you need to absolutely avoid is commonly called "snuffles" by rabbit raisers and will be called pasteurellosis, rhinitis, or pneumonia by your veterinarian. This disease is the reason we recommend you visit your breeder *before* purchasing your rabbit and listen for sneezing,

look for clear eyes, and make sure there are no runny noses or matted fur on the insides of the front feet (indicating cleaning of runny noses). Unfortunately, the breeder may have carrier rabbits that they may not even be aware of that show no clinical signs of disease in his rabbitry. There are varying estimates that between 30 percent to as many as 70 percent of apparently healthy rabbits may be carriers of *Pasteurella multocida,* the bacteria that causes snuffles. It is actually preferable that, if an animal is a carrier, it be discovered during the quarantine time and not after you have introduced the animal into your operation. Why is this? What's so bad about this disease?

First of all, *Pasteurella* bacteria are highly contagious and transmitted by direct contact or by sneezing—so it can spread very rapidly through a barn. The most common clinical signs of pasteurellosis include rhinitis (runny or stuffy

Upper respiratory infections can be fatal to the rabbit AND to the rabbit ranching enterprise.

Wry neck in rabbits can be caused by several organisms, including *Pasteurella*. It is considered fatal.

nose) and pneumonia (inflammation of the lungs). Rabbits are obligatory nasal breathers due to their anatomy. Therefore, a stuffy nose in a rabbit is far more serious than for a human who can switch to mouth breathing if necessary. Rabbits that develop pneumonia rarely recover, and if they do, they are still carriers of the *Pasteurella* bacteria.

Wry neck or head tilt (which is just that—a twisted neck that the rabbit can't straighten, making it difficult or impossible to walk, eat, and drink) can be caused by various ear disorders, a protozoan parasite (*E. cuniculi*), or *Pasteurella* infection in the ear, brain, or nervous system. Wry neck in rabbits is considered terminal.

An adult female or male with *Pasteurella* infection of the genitalia can have a puslike discharge and become sterile (after they have passed it to any other animal with whom they have been bred). Note that other bacteria can cause these symptoms, but *Pasteurella* is by far the most common.

Upper respiratory disease (rhinitis) usually appears first if an animal is harboring the bacteria. Then, stress, poor ventilation, unsanitary conditions, and dirty nesting material contribute to the escalation of the disease to pneumonia or other problems.

So what do you do if your new rabbit in quarantine starts showing symptoms? It sounds heartless, but we recommend euthanasia. This is preferable to the dozens or hundreds of animals that might become victims in the future. We are not talking, of course, about a single sneeze or two in the three-week quarantine—like any animal, a rabbit may sneeze from dust or an irritant, but frequent sneezing, matted fur on front legs, or weepy eyes with the sneezing are warning signs of possible *Pasteurella* infection. If you are unsure, make a trip to your veterinarian for diagnosis. Signs of an active *Pasteurella* infection in a rabbit during the stress of transport and quarantine is an indication that it has a poor immune system that you do not want to perpetuate in your breeding stock. We understand that with the rather large percentage of rabbits that harbor this bacteria, some people may be horrified by the suggestion of euthanasia. But we are not talking here about a single pet rabbit that can be isolated and treated. We are giving advice to the commercial rabbit raiser. We can not stress enough the economic importance of this disease and the heartbreak it can bring to a rabbit ranch.

As will be mentioned in the housing section of this book, proper ventilation and stress-free, sanitary conditions are an absolute necessity to prevent outbreaks of pasteurellosis in a rabbitry. Some research has even indicated a direct correlation between cases of snuffles and high ammonia levels in rabbit facilities (i.e., improper ventilation).

DIARRHEA

A rabbit should have large, firm, moist droppings. Diarrhea may occur briefly with a diet change or stress but should resolve on its own. If a new rabbit has staining around the anus or diarrhea after you have had them a few days, you should take a sample to your veterinarian and have them check for coccidia (a common protozoan parasite with the scientific name of *Eimeria*) or other intestinal parasites before you consider introducing the new rabbit to your facility. If you have purchased a juvenile rabbit under four months of age and they have diarrhea or stomach bloating for any reason, you may have a serious condition that needs veterinary care immediately. Diarrhea in a young rabbit can kill in under 24 hours.

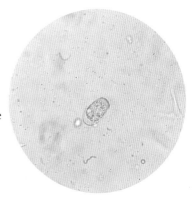

Coccidia oocysts can survive in the environment for many months.

Most *Eimeria* species inhabit the intestine, and one inhabits the liver. Coccidiosis is the term for the disease state resulting from a severe infestation of *Eimeria,* and along with diarrhea, there is weakness, dehydration, weight loss, abdominal pain and bloating, and death. The liver form will be visible as spots on the liver and result in rejection of the rabbit meat by most inspectors, but the intestinal form is more often fatal. Low levels of coccidia may be present without causing any symptoms, but stress, overcrowding, and unsanitary conditions can cause proliferation of the *Eimeria* to the point where it causes disease in the animal.

Coccidiosis in your kits can be controlled by giving them a coccidiostat to kill the parasite, but this is often not desirable in a food animal by your customers. Many meatpacking plants will not buy rabbits that have been given coccidiostats. It is far better to work with your veterinarian to eliminate the parasite, enhance your sanitation, and improve your kits' immune systems rather than treat the parasite itself. Coccidiosis is avoidable but takes vigilance. Using 10 percent ammonia when cleaning cages and the use of heat torches on cage surfaces (with animals removed first, of course) can aid in killing the coccidia oocysts that are passed in the feces and spread the disease, and which can otherwise last for up to a year in the environment.

If your rabbit has diarrhea and the veterinarian can find no evidence of coccidia, they will usually call it enteritis, which is a generic term for inflammation of the intestine. Often, the cause is unknown. Many times, it is the result of feeding "treats" or not feeding enough hay (see Chapter 4), or it could be caused by bacteria from contaminated feed or water or from allowing the rabbit to eat wilted greens.

Stress is an often overlooked cause of enteritis, especially in kits. Adrenaline surges from stressful

Watery diarrhea, often observed as wet legs and hindquarters, is especially devastating in kits, which can then die in under 24 hours. Along with "snuffles," it is one of the most economically significant ailments of the commercial rabbitry.

situations slows the food passage in the gut and changes the pH, allowing for the growth of undesirable pathogenic bacteria. It may take up to a week from the stress event to the occurrence of diarrhea. Stresses can be from the presence of predators, rowdy people, loud noises, or from excessive handling, removing the safety of their nest box, weaning, or simply moving to a new cage. Though not all of these can be avoided, combinations of them increase the likelihood of stress-induced diarrhea.

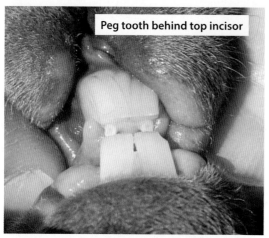

Peg tooth behind top incisor

A rabbit has two small peg teeth directly behind the top incisors. In a normal dental alignment, the bottom two incisors will rest tight behind the top incisors on the peg teeth.

Note the sharp, chisel-shaped incisors for tearing the normal rabbit's diet of grass and other vegetation.

DENTAL PROBLEMS IN RABBITS

Rabbits have four top incisors (front teeth). The two most visible are large and chisel-shaped, with two small, tubular-shaped peg teeth behind them (see photo). They also have two lower incisors, which should line up so that they rest on the upper peg teeth. The incisors are used for tearing and cutting the food into manageable pieces. Rabbits also have molars, or cheek teeth, to grind their food.

Malocclusion

A rabbit's teeth grow continually throughout their life, and if there is any misalignment—or as a veterinarian would say, *malocclusion*—they are not worn down evenly and will quickly cause problems such as oral abscesses, difficulty eating, weight loss, and lethargy.

A rabbit with malocclusion may drool and not want to groom itself. Checking for properly aligned teeth should be part of the health inspection, both when you first acquire a new rabbit and on a routine basis, especially since the teeth are constantly growing and changing. It is difficult to examine a rabbit's cheek teeth without sedation, but misaligned molars will eventually result in front incisors that do not wear properly.

If a young rabbit has malocclusion discovered in the quarantine period, you should reconsider using them as a breeder. Just as in humans, some rabbits are born with a hereditary misalignment of their teeth. This is termed *congenital malocclusion* and is not something you want to introduce into your breeding program. However, please note that a rabbit can also suffer from *acquired malocclusion* at later stages in life.

One contributing factor to acquired malocclusion is when the rabbit is fed only soft commercial pellets. Theoretically, perfectly aligned rabbit teeth should wear down on each other, but their constant growth means even a slight malocclusion can worsen rapidly without the normal roughage a rabbit in the wild can obtain. As will be seen in Chapter 4, hay is recommended for rabbits for myriad reasons, and dental health is one.

Chew Sticks

We always provided chew sticks in addition to hay and rarely saw dental issues. Be aware that while some rabbits will have a blast playing with their stick toys, others seem to consider them a personal invasion of their space and will do their best to shove them back out of the cage.

Chew sticks are the one exception to wood in a rabbit cage. Wood is porous and impossible to disinfect, but a chew stick is only there for a short time before it is replaced with a fresh one. Do not leave sticks contaminated with feces or urine in the cage. Also, never take a stick from one cage and give it to another rabbit.

It could be argued that it is an unnecessary risk to provide sticks that may have been contaminated by wild animals or insects to your rabbits. But remember that your hay, and even pellets, carry the same risk. No facility is completely biosecure. Still, to reduce worry, we suggest sticks from upright trees, not old dead decaying sticks found on the ground.

Vitamin Deficiency

There is also a metabolic form of acquired dental disease that is usually diet related. Improper balance or inadequate levels of calcium and phosphorous or vitamin D deficiency (which regulates calcium/phosphorous absorption) can lead to demineralization of the bones. This can lead to irregular or misaligned teeth. As in humans, vitamin D is synthesized in the rabbit in

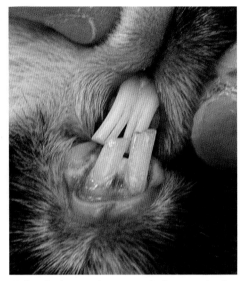

Malocclusion can be congenital or acquired. The acquired form is often the result of failure to provide hay or other roughage to help wear down the continually growing teeth in rabbits.

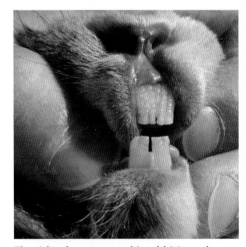

The ridged pattern on this rabbit's teeth was from a metabolic disease that affected the teeth. Abnormal teeth, especially in multiple rabbits, indicates you may need to reevaluate your feed.

response to sunlight, but it can also be provided in the diet. Most pelleted feeds, along with quality hay, will provide adequate levels of these substances. Anytime you begin to observe abnormal teeth, especially in multiple rabbits, it is time to evaluate your feed with your veterinarian.

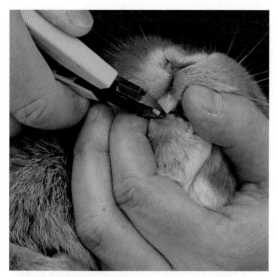

NEVER attempt rabbit dentistry at home. It is a job for an experienced veterinarian. Prevent rather than correct the problem!

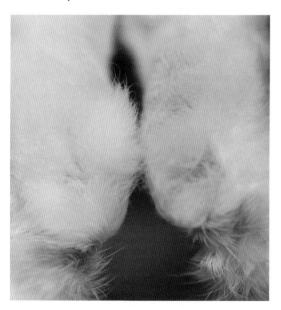

The early stages of sore hock show missing fur and slightly reddened or thickened skin. This is the time to make the husbandry changes that will prevent the skin condition from becoming worse.

How to Fix the Problem

What do you do if you have rabbits with malocclusion or other teeth abnormalities? First, find out what is causing the problem and eliminate it from your rabbitry—whether it is a genetic predisposition, failure on your part to provide the necessary roughage to chew, or a metabolic disease from an improper diet. If the rabbit is very valuable to your breeding program, and you want to correct the dental issue, take it to your veterinarian for treatment! Do not attempt clipping the teeth with wire cutters or similar tools. The animal needs to be sedated and the proper tools utilized to accomplish dental correction. The angles of the teeth must be exact to prevent further problems. It is also very easy to break off or crack a tooth, which will lead to even more pain and abscesses.

SORE HOCK AND OTHER SKIN DISEASES

Whole books have been written about skin disease in rabbits. Again (not to be too repetitious), prevent, prevent, prevent. While your new rabbit is in quarantine, check it closely for signs of skin rash, missing fur, excessive itching, or any lumps, bumps, or visible ectoparasites. If any

TIP

If you're noticing the development of sore hock in your rabbits, there are a few things you can do.
- Check that the cage floor provides comfortable footing.
- Inspect the rabbit's toenail length.
- Determine if they are overweight.
- Change to a different resting board.
- Most importantly, thoroughly disinfect the cage and look for leaking water sources.

of this occurs, call your veterinarian and get it diagnosed and treated before introducing your new rabbit to your rabbitry. Most skin conditions are contagious, and like ear mites, if your rabbits do not have fleas, fur mites, mange mites, lice, or ringworm, you certainly do not want to introduce them!

There is one very common skin problem that only affects the feet, It's not contagious, but it is often the result of improper rabbit husbandry. The medical name is *ulcerative pododermatitis,* but it is more commonly known as sore hock. It is a pressure sore or ulceration of the footpad.

Hopefully, a newly purchased animal in quarantine does not have this problem, but it can develop at any time. Every rabbit in your facility should be checked at least weekly to catch it in an early stage. It is more common in heavier breeds, and obesity is a major risk factor. There

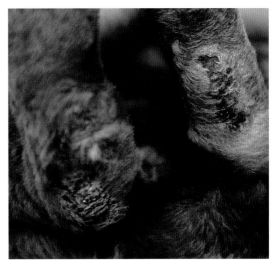

The later stages of sore hock are extremely painful, difficult to treat, and can lead to death.

is some genetic predisposition among individuals who have thinner footpad skin or hair, or slight musculoskeletal irregularities, which place the weight too far back on the hock. Other factors that can lead to the condition are rough or rusted flooring, flooring that is too smooth to allow a proper foothold, chronically damp conditions, insufficient room to move around or fully stretch out, overgrown toenails, or lack of an alternative resting board in wire-floor cages (see Chapter 5).

Sore hock in the early stages may only be evidenced by missing fur on the footpad and some thickening or redness of the skin. It can start in one foot, but then a shifting of weight off of the sore spot can lead to pressure sores on other feet as well. Later stages manifest as open sores, bleeding, infection, and if left untreated, can result in death. When a suspected case of sore hock is developing, immediately begin an evaluation as to the possible causes.

If caught early, this condition is treatable by the mentioned husbandry changes and with topical antibiotics as prescribed by your veterinarian. If you find a particular line of rabbits that seems especially susceptible, that line should be eliminated from your breeding program.

MYXOMATOSIS, TULAREMIA, AND RHVD

Though all three of these diseases are still very rare in the United States, incidences have increased, and there are preventative measures that can be employed to lessen your risk. Myxomatosis is not common (yet) in the United States and has only been identified on the West coast. Still, it has so devastated the rabbit populations of other countries that it should be mentioned. It is a poxvirus that causes a very mild disease in its original wild rabbit host in South America (and the wild host rabbits in California), but it is almost always fatal to rabbits descended from European domestic stock. This disease was actually deliberately introduced to France and Australia to control wild rabbit populations there in the 1950s and

Myxomatosis is rare but present in some parts of the United States

has since spread throughout Europe. It is transmitted by direct contact or by biting insects such as mosquitos, flies, fleas, or ticks. Rabbits will be lethargic, go off food, and have fever, skin lesions or nodules (usually on the face, ears, and eyelids), respiratory infections, and pneumonia. Rabbit raisers on the West coast should consider investing in mosquito screens for their rabbitries if cases there continue to increase.

Tularemia (also known as rabbit fever) is another very rare disease present in many areas of the United States that is also spread predominately by ticks and other biting insects, or by direct contact with an infected animal. Other than rabbits and hares, rodents are a major reservoir species for the bacterium *Francisella tularensis*, which causes this disease. The symptoms are similar to myxomatosis (weakness and fever, ulcers, and pneumonia), and without antibiotic treatment, it results in death. Humans can contract tularemia, which is why rabbit hunters are careful to use gloves while skinning wild rabbits. Housing animals off the ground with strict attention to insect and rodent control are the primary methods to prevent this disease.

Rabbit Hemorrhagic Virus Disease (abbreviated RHVD, RHD, or RHV) is a highly contagious disease caused by a very hardy calicivirus that can survive in the environment for months. It is rare in the United States but has very recently been found in both our native and domestic rabbits, and so may be spreading. It is not a health risk to humans, but in rabbits, RHVD is a sudden killer. If symptoms are noted before death, they are usually:

- High fever
- Seizures
- Difficulty breathing
- Bleeding from the nose, mouth, or rectum

RHVD is spread by direct contact with an infected rabbit, or by contact with objects, people, or other animal species that have come in contact with the virus. Insects, such as biting flies, fleas, and mosquitos, can also spread it.

Housing off the ground and insect and rodent control are key to preventing all three of the rare but devastating diseases mentioned above. All three are also "reportable diseases" in most states, meaning that once diagnosed, federal and state agricultural or health authorities may need to be informed.

PREVENT, DON'T CURE, DISEASES

The most vital takeaway from this section is the importance of prevention. Any time you introduce a new rabbit, you risk introducing disease. If your operation has hundreds of rabbits, this could be a costly and terminal mistake. The more you can raise your own replacement breeders, the safer your operation is.

We recommend that you begin with enough does and bucks to reach the level of production you want in the first year or two. Remember to purchase the rabbits from one or two reputable commercial breeders after you have thoroughly checked out and toured their facilities. Eventually, you may need to add new blood for genetic considerations. Hopefully, this can be done with only the addition of a few exceptional bucks to reduce your risk.

Please do not let the possible rabbit diseases mentioned here deter you from raising rabbits. Farming any livestock takes education and vigilance to keep them healthy. With proper husbandry and preventative care, you may never see any of the illnesses mentioned above, especially if you adhere to the following rules.

- Contact a veterinarian regarding any suspect animals before they are introduced.
- Make regular weekly health checks of every rabbit from "stem to stern."
- Disinfect cages and equipment religiously.
- Never keep animals in overcrowded conditions.
- Provide adequate ventilation in barns.
- Get serious with insect and rodent control.
- Eliminate stress factors as much as possible.
- Avoid sudden dietary changes and feed the rabbits quality food appropriate to their life stage.
- Provide fresh water at all times in a sanitary manner.
- Do not use animals that fall ill (even if they recover) as brood stock.
- Always carefully consider before you take rabbits to fairs or shows where they have close contact with other animals. If you do take them along, the animal should be returned to a quarantine area just as if it were a new animal.
- Do not be afraid to ask other rabbit raisers to wear disposable gloves and smocks when they are touring your facility. If they are serious rabbit farmers, they should appreciate your bio-security measures.

Any rabbit that leaves the farm for any reason, such as participating in a show or fair, should be quarantined when returning, just as if they were a new rabbit.

CHAPTER 3

BREEDING, PREGNANCY, AND BIRTHING

THE BREEDING ACT

Breeding is the heart of a meat rabbit facility. How to breed, when to begin breeding, how often to breed, who to breed with whom—all these factors contribute to the success or failure of an operation.

The breeding act itself is quite simple and very fast. Always take the doe to the buck's cage, not vice versa, for two reasons:

1. The buck is comfortable in his cage and has scent-marked it. He won't waste time trying to mark a new territory. The male scent will also help trigger the doe to "lift" her hindquarters and allow the buck to complete ejaculation.

2. The female tends to be protective of her cage territory and may be aggressive to the buck if brought to her cage, possibly even injuring him.

During the actual mating act, the buck will chase the female around his cage a bit, mount her from behind, and hold her still with his teeth or forepaws. When she lifts her hindquarters, he can complete ejaculation. He will usually make a grunting or squealing sound during ejaculation and fall over on his side. This all happens—if it is going to—in less than ten minutes. You should never leave the female in the male's cage unattended during this time. If a male is trying to aggressively breed an unreceptive female, she may injure him severely, even biting his testicles. Also, since the mating act is fast, if you walk off for even a few minutes, you may not know she is bred and not give her a nest box at the appropriate time.

The male is nearly always ready to breed if he is healthy and in good condition. This is if he is used for a reasonable amount of time. Five or six days a week, once a day, has been shown not to affect sperm count or viability. An older buck that is hot or an obese buck may be less enthusiastic about breeding. In fact, any time the heat in the rabbitry is over ninety degrees, attempting to breed may heat-stress the rabbits and will rarely result in successful breeding.

> **Rabbit ovulation occurs *after* sexual contact.**

It is also necessary that the doe is in a receptive stage hormonally for success in breeding. The estrus cycle of most other domestic mammals takes place at regular intervals. Estrus is when the female is in heat and ovulation (eggs being released) occurs. Rabbits, however, do not have a set heat cycle like dogs, sheep, goats, pigs, horses, or cattle. That is, they do not come into a "season" or specific time period when they are receptive to the male. One of the reasons for this is that other animals put a lot of their body's resources into months of pregnancy and then nursing and nurturing their young for months more. Thus, it is imperative that the young are born at optimal times of the year for successful rearing, and their limited heat cycle is to help confine them to these specific times.

The rabbit, in contrast, is a prey animal at the bottom of the food chain. They have short pregnancies and short nursing stages (about one month for each). In the wild, they are naturally geared toward getting pregnant and passing on their genes as soon as possible to as many offspring as possible—before they are somebody's dinner. Therefore, they do not have an estrus cycle with regular periods of heat in which ovulation will occur. Instead, rabbit ovulation occurs *after* sexual contact. This is called *induced*

A doe needs to be in a receptive state before breeding. Otherwise, chances for successful conception are low.

ovulation. Any eggs in the ovary that are not ovulated through lack of sexual stimulation are simply reabsorbed and replaced by new eggs periodically. These new eggs remain for a few days in the preovulating state and then, in turn, regress if no mating occurs. This cycle continues throughout the doe rabbit's reproductive life.

This does not mean that a doe will breed at *any* time. A female rabbit is considered to be in "heat" not when she is ovulating, since this occurs after mating, but when she accepts service from the male. It should be noted that 90 percent of the time, when a doe has a swollen red vulva, she will accept mating and then ovulate. However, the amount of color change and swelling varies greatly with the individual doe. Also, she may mate if she is just coming into her receptive state and not yet showing visible physical changes to the vulva.

A doe in heat may run briefly from an overeager male (or may chase him instead), but she soon assumes a characteristic pose where her hindquarters will be raised to assist

A litter of two-week-old Flemish Giant kits.

in mating. A doe not in heat tends to crouch with her tail backed up into the corner of the cage or even exhibit aggression toward the buck. The sexual behavior of a female rabbit is thus very special. She has no set heat cycle but will be sexually receptive and assist the male during mating between 80 and 90 percent of the time.

You may ask: "Should I wait to breed a doe until I see these actual physical changes in the vulva?" The answer will depend on your individual operation and the time you have to spend breeding. Due to market concerns, which will be discussed later, a meat rabbit facility often wants a large group of rabbits all born at the same time so that they can be taken to market as a group, and also so that extra kits can be easily fostered between different moms. Thus, the rabbit breeder is usually trying to get a set number of does all bred within a few days of each other. If you only have a few bucks, yes, it would be best to check each doe and find the most receptive ones (with a dark red swollen vulva) to breed for that day, rather than wear out the poor buck by putting several nonreceptive does in with him. If you have excess bucks, you can try does that are not showing vulva changes, and if they don't breed, they can be tried again the next day. The very act of being put in with a male (and his scent-marking pheromones) may trigger her receptivity.

The female rabbit is unique in another way. In most other mammals, the progesterone secreted during gestation (pregnancy) immediately inhibits any further mating. However, a pregnant rabbit is one of the few animals that may accept mating throughout the gestation period. This means two things for the rabbit breeder:

1. You cannot use the sexual behavior of a doe as a definitive indication of pregnancy. She may be pregnant and still mate again if put in with a male.
2. The rabbit doe can be bred more than once to ensure fertilization of as many eggs as possible.

It has been shown that mating occurring during very early gestation has no dire consequences for the embryos. Some farmers will breed a doe in the morning and then in the evening of the same day, and others will breed two days in a row to ensure as many sperm meet as many eggs as possible. Breeding in the morning and then again in the evening makes perfect sense, as the eggs will be moving down from the ovary when new sperm is introduced. Due to time constraints, our favorite method is to leave the doe with the male during her initial mating long enough for him to breed her at least twice (preferably three times). This method always ensures good-size litters in our rabbitry.

AGE TO BREED

The age the rabbits are first bred is critical to your operation as well. Male meat rabbits such as New Zealand Whites develop fully mature sperm by 135–140 days old. The majority are psychologically ready to breed by this age, as well. You should, however, make sure that your virgin male's first mating attempt is successful by putting him in with a young, eager female that is at a receptive stage. You do not want to put him in with an older female who is not currently receptive, as she may injure or scare him and actually make him afraid of the mating act.

The optimal first breeding of the female is between 4.5 and 5 months of age.

The female meat rabbit has been studied extensively. Most experts agree that the optimal first breeding of the female is between 4.5 and 5 months of age. (This is why we recommended in the preceding chapter that your rabbits be purchased at four months of age, allowing for quarantine before breeding.) At over six months of age, some females may become overweight if they are not in a breeding program, which results in decreased productivity or even failure to breed. Rabbits tend to put on fat internally rather than as an external layer, as in some animals. Excess fat around the ovaries will often prevent conception.

TEMPERATURE AND LIGHT IN RELATION TO BREEDING SUCCESS

Rabbits may experience heat stress by attempts to breed them when it is too hot. It has also been reported that the rabbit buck may be temporarily sterile in temperatures over 90°F due to decreased sperm motility and viability.

So what do you do if you live in an area of the country with long stretches of over 90°F weather? There are several options:

1. Keep close tabs on the long-term weather predictions and plan breeding days for a stretch of relatively cooler weather.
2. Breed early in the morning when it is cooler.

3. Keep your bucks in an air-conditioned facility during the summer. If air conditioning is not an option, you can provide them with fans to keep their cages cooler and plastic soda bottles filled with water and frozen.

The length of time that light is provided is also an important and often-overlooked factor in rabbit breeding. Experts have found that male rabbits have slightly more sperm in their ejaculates with 16 rather than 8 hours of light. It is well established that the female rabbit is far more opposed to mating if only given 8 hours of light a day. It is therefore recommended that your rabbitry be lighted for 15–17 hours a day for optimal breeding success. Full-spectrum lighting is best. If you have a facility without electricity, you will have to expect decreased litters in the short-daylight months and a lot more time spent in taking the females to the males. In this case, save good breeding stock from animals that will breed in the fall and winter so you do not end up with a period of time when your rabbitry is not in production.

PALPATING DOES FOR PREGNANCY (OR NOT)?

There is some controversy over whether or not to palpate your rabbits to check for pregnancy. To learn to do it, get an experienced rabbit breeder to show you how, or at least watch several YouTube videos. Our question is "Will the time you need to spend and stress for the animal outweigh the benefit of knowing if the doe is pregnant?" A rabbit fetus grows very slowly for the first three weeks and then very rapidly for the final week of gestation. Most experts recommend that palpation should be at 10–14 days. After that time, it is a greater risk to the developing kit. At 10–14 days, the fetus is only about the size of a pea or marble (which may be hard to locate). You must determine how the knowledge of pregnancy two weeks before birth will affect your operation. Would it change your actions with the doe? Obviously, if you determine that the female is not pregnant, she can be rebred and save those two weeks of non-productivity. The deciding question is *if* you can determine for sure the doe is pregnant using palpation.

At Chigger Ridge Ranch, we *never* palpated our does for the following reasons:

1. If you feel nothing during palpation, it could be the result of your palpation technique, how tensed the rabbit is, or how many kits are present. So the question then is "Do you try to rebreed her after feeling nothing or wait two more weeks?" Our answer is wait. Remember that a doe may accept a buck even if she is pregnant, unlike most other animals—so that would be no indication that she wasn't pregnant from the initial breeding. Having her running around to avoid the buck if she is pregnant and not receptive is a risk for the fetal kits.

2. If you do feel something during palpation, you still aren't 100 percent positive and will simply be waiting to be sure in two more weeks.

3. There is always a possibility of injury to the doe or developing kits.

4. It is an additional (and, in our opinion, unnecessary) stress for a doe to be restrained and have her abdomen squeezed.

5. It is very time-consuming for the rabbit raiser if you are trying to palpate dozens of does every couple of weeks. We found in our operation that a doe that allowed a buck to mount her at least twice during the mating procedure and lifted her hindquarters to assist him had a 95 percent chance or greater of getting pregnant. Therefore, the time and stress on the animals for palpation was not considered advantageous.

FEEDING DURING PREGNANCY

The feeding schedule should not change during pregnancy. Rabbit fetuses are very small for the first 21 days (grape size or less). A doe fed additional feed during this time may actually put on too much fat.

From day 21–31, the fetuses begin to grow rapidly. However, they are still relatively small until the last few days of gestation. At that time, you might want to increase food, but this is also the point where the doe generally does not want to eat as much as she prepares to kindle (have her kits). She may even go completely off feed for the last 24–48 hours. However, she will usually be starving immediately after kindling and will need additional feed to prevent her body from metabolizing all her stored fat to provide the added energy required at this time.

At Chigger Ridge Rabbits, we fed the normal ration of 4–5 ounces of 18 percent protein rabbit pellets and free-choice quality hay until the day of birth. On seeing live kits, the doe was given a double ration (8–10 ounces) of pellets. We then moved into our lactation feeding schedule (see Chapter 4). Lactation takes much more out of a doe than pregnancy.

The accompanying graph gives a visual idea as to the growth rate of the domestic rabbit fetus. These are average weights taken from a multitude of scientific studies conducted for various reasons over the years. These growth rates will vary with the doe rabbit's breed and condition. This chart is intended only to give an idea of how slowly the fetus grows and how small it is until day 21, at which time the very rapid increase in weight begins.

Do NOT feed a doe extra rations during early gestation, as this can lead to obesity and complications during birth.

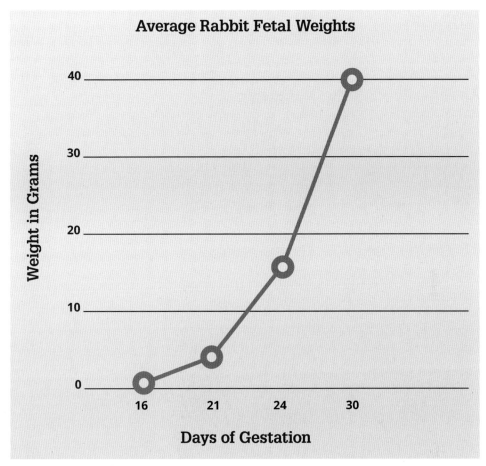

Average Rabbit Fetal Weights

Weight in Grams

Days of Gestation

Rabbit fetuses are only about the size of a marble until 21 days, then they increase rapidly in weight for the last 10 days of gestation.

If the kits die in the uterus and are under 21 days, they are small enough that they are generally reabsorbed by the mother, and you will not even know there was a pregnancy. If they die over 21 days of age when they are larger, the doe will abort them, and you will see some blood or tissue in the cage. Abortions are rare in our experience and usually the result of some major systemic stress, illness, or trauma.

It is unusual for a doe to be unable to deliver her kits. There are two scenarios where this is more likely to occur:

1. There is only a single kit or two that have grown too large for the doe to pass easily.
2. The doe herself is obese.

Both circumstances offer further reasons *NOT* to feed the doe extra during pregnancy. We had people constantly tell us they feed does extra "just in case." Unless the doe was too thin when she was bred, you are doing much more harm than good for her. Also, if she turns out not to be pregnant, you have increased the chances that she will not conceive at her next breeding due to obesity.

NEST BOXES

There are basically three different options for nest boxes in a standard rabbit hutch. One is a box that sits on the floor of the cage, and the doe must jump up to hop over the sides to get in and out of it. Another drops down below the floor of the cage and is constructed into the cage bottom so that the rabbit hops down into it. The third option is a section of cage that is partitioned off and only opened when the rabbit has kits. The second is the most natural for the doe, as in the wild she would be digging a hole or depression in the ground and having her kits down below the ground. Any of these systems will work, however, as long as the doe feels secure in the nest.

A more important consideration than location of the box is the material from which it is made. We do not recommend solid metal boxes. They are hotter in summer and colder in winter. In fact, in cold weather, if a kit burrows down under the nesting material and lays on the cold metal, the kit may become hypothermic (body temperature too low) and die. We also do not recommend using solid wood nest boxes. Although they are cheap and easy to make, wood is impossible to clean adequately. A doe can get mastitis in her mammary glands or get a uterine infection, or there may be a dead kit hidden under the nesting material undetected for a time—all circumstances that require sterilization of the nest before it can be used again. With wood, this is simply not possible—its porous nature allows bacteria to hide, no matter how much it is scrubbed and bleached. Another factor to consider when selecting the material for nesting boxes is that female rabbits tend to mark their nests by defecating and sometimes urinating in them. This means the nest should have some sort of draining ability.

A commercial system for nesting is often a "hole" that can be opened to a separate portion of the cage when a doe is ready to kindle.

Universal Sani-Nests have disposable cardboard liners to ensure a hygienic first few weeks, and they are big enough to accommodate New Zealand or Californian does with full-size litters.

In our rabbitry, we finally went with Bass Equipment Company's Universal Sani-Nest. These are wire nest boxes 18 inches long by 10 inches wide by 8 inches high. This is a good size for the large commercial New Zealand rabbits. Smaller boxes are too cramped, leading to trampled kits or kits born out of the box. Larger boxes take up too much of the doe's living space. The wire of the Sani-Nests are 1 inch by ½ inch. These wire nests are then lined with cardboard. This prevents the industrious mom from pushing all the nesting material out of the wire during nest building.

A person can make these wire boxes if they have time, but if you do, be sure to bend the wire at the top over in a manner similar to Bass's model to prevent the doe from snagging a foot or teat on sharp wire. Cardboard inserts for the Sani-Nests are also sold by the Bass Equipment Company. Or you can cut out your own liners from scrap cardboard. For those of you concerned with sustainable agriculture, the used nest box liners work great under rabbit cages to protect your worms from rabbit urine and give them cover and retained moisture.

Nest boxes should be placed with the doe 28 days after breeding. You must provide some sort of nesting material in the box. The doe will pull fur from her own body to cover her kits, but she needs some straw or shavings to give her a base for her nest. What you give her to form her nest is dictated by the area where you live. As long as it is dust free, dry, absorbent, and clean, go with whatever is cheapest in your locale. Give her enough, even in warm weather, so the nest is thick enough that she will not squash kits when she jumps into and out of the nest. Kits have almost rubberlike bones when born (which is why they can survive even a 4-foot drop to the hard floor if they crawl out of the cage).

Still, mom standing on them on a hard, unyielding surface may result in scratches or prevent them from breathing. The kits need to be able to wriggle away under the nesting material.

Most does will kindle (have kits) at 31 days, but depending on litter size, condition of doe, and individual physiology, it may be as soon as 29 or as late as 33 days. We always leave the box in till day 34 to be absolutely sure we didn't simply make a breeding date recording error. You do not want to put the box in before day 28, however, as she may foul it.

Along those lines, do not put the box in the corner of the cage that the rabbit has chosen for a "toilet." Rabbits generally choose one spot in the cage for defecation and urination. Some does will change their bathroom habits if a nest box suddenly appears there; others will continue to use their chosen corner no matter what! As mentioned, a doe may defecate or put a few drops of urine in a nest to mark it, but daily defecation and urination in the nest obviously does not contribute to healthy kits.

It is normal for a doe to remove nesting material and wander around the cage with it in her mouth before kindling. However, if she begins to place it in a specific spot out of the nest and adds more to it, she has decided that this is where she is going to have her babies, and nothing will change her mind. Move the nest to that location or she will likely kindle out of the nest. This is another reason we chose to use the Sani-Nests rather than a more built-in nesting system.

Newborn kit rabbits are born hairless and with eyes closed.

BIRTHING, FOSTERING, AND HAND REARING

Birthing is the easiest of any topic to write on. It is simple: There are only three rules:

Rule #1: **Leave them alone.**

Rule #2: **Don't bother them.**

Rule #3: Let them bond with their kits, **undisturbed.**

The object is to let the doe feel secure and comfortable by reducing any noise or disturbance for at least 24 hours before, during, and following kindling as she is bonding with her kits. A new mother may cannibalize her newborn kits if disturbed too much. This happens more often with first-time moms or naturally nervous does, but the presence of predators or loud kids may trigger it. (A buck housed right next to a doe may also trigger it.) We wait a full 24–36 hours before disturbing the doe's nest at all. If there is a dead kit that the doe doesn't consume, this amount of time is not enough to affect the health of the other kits. If you see movement under the fur the mother has pulled from herself to cover her kits, that is all you need to know for the first 24–36 hours.

While we are on the subject of fur pulling: some does pull an excessive amount of fur and pile it on 6 inches deep, even at the height of summer. Others are content with a token mouthful or two. If it is summer, we will often take the extra fur and store it for the rare occasion when a doe hasn't pulled enough fur to keep the kits from becoming hypothermic in cold winters. We have found nothing else that beats natural rabbit fur for nest warmth, and never found a rabbit mom that minded another doe's fur being added to her nest if needed. That is the only exception to the "do not disturb" rule. If it is cold weather and there is not much fur in the nest, adding fur is acceptable. Nothing will kill a kit quicker than hypothermia.

Holding the kit in this position while feeding may lead to aspiration of milk into the lungs.

At 24–36 hours, it is time to peek under the fur and remove any dead kits or placenta that the doe has not already consumed, and replace soiled nesting material. It is a perfectly natural behavior for the doe to consume dead newborn kits and otherwise clean her nest of anything that might attract predators. If you do need to clean up the nest some more, the easiest way to do this is to have another container ready to which you quickly transfer the kits with the clean nesting material and all the dry fur. Remove any damp material from the nest and add additional nesting material if needed, then replace kits and fur and return the nest box to the mom's cage (this should be done once a week until the litter is weaned).

When you first check the nest, you will also want to count the kits, and if necessary, you can foster extra kits with other moms. Moms that have more than eight kits can be relieved of the extra burden by placing their excess kits with moms that have less than eight. (The reason this number is chosen is that a doe has eight teats.)

There is no trick whatsoever to fostering kits. Simply take the extra kits and place them with the new mom. Choose the strongest, fattest kits to transfer if the litter you are moving them to is a day or so older, and pick smaller kits if you are transferring them to a younger litter. Rabbit females, unlike dogs or cats, never pick up their young. A newcomer in the nest is pretty much guaranteed acceptance. The only way a mom can reject them is to reject the entire litter, which she will not do.

If necessary, you can foster extra kits to other moms.

There are two things to consider when fostering, however. The first is whether or not both moms are healthy and doing well. If a doe has mastitis (infected teat) or any other diseases, the fostered kit may transfer the infection to the healthy doe and litter. The second consideration is whether or not you might want to keep a future replacement kit for yourself (or to sell) from the litter. Once you have added a foster kit to a litter, you will not be able to tell it apart from the others. You can mark a fostered kit with a touch of tattoo ink placed on an ear, although this may have to be renewed every so often.

What do you do if all of the does have eight or more kits and there are no moms with small litters available to take on foster babies? This is always a difficult question. A good, heavy-milking New Zealand or Californian can successfully raise a very nice litter of up to 10 kits without too much stress on her or missed meals for the kits. However, a doe rabbit has eight teats and only feeds her kits one or two times a day. So the more kits in a litter, the more will have missed meals. With litters of over 10 kits, it becomes obvious that either some will slowly starve to death or the entire litter will be stunted in growth. It is usually the former. This leaves the rabbit breeder with three choices:

1. Let "nature take its course" and let some kits starve.
2. Euthanize the weaker kits rather than let them starve (incidentally, they can be frozen and sold to pet stores as snake food).
3. Hand rear a few kits.

Hand rearing is not all that difficult if the kits received at least some of the "first milk," known as colostrum, from the mom. This is the milk that comes in the first 24 hours and is full of the antibodies that protect the kit from disease for its first 2–3 months of life until it can produce its own antibodies. If all kits got a minimum dose of the colostrum, raising them is fairly easy (unfortunately, there is no way to tell if the kit has managed to make it to that first all-important meal).

The rabbit doe only nurses her kits once or twice a day, so it is fairly easy to work hand rearing kits into a normal human's workday (unlike some animals that nurse every 2–3 hours). With a little experience, it only takes 15–20 minutes to feed a whole litter. Kitten milk replacer (KMR), which comes in cans as a liquid and is sold at veterinary clinics or pet stores, and lamb milk replacer, which is a powdered milk that you mix with water (available at most farmers co-ops or feed stores), are the closest to rabbit milk available. What you need is the highest-possible fat content. Lamb milk replacer is usually

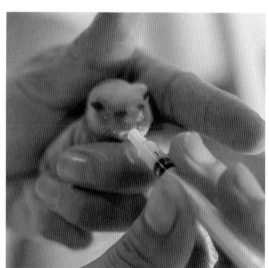

much cheaper in general than KMR. You can use pet nurser bottles, but we found that the rabbit kit is very quick to destroy these after a few days. Syringes (without needles) work fine and are much less liable to damage. Feed the kit as much as it will take at each feeding. Remember, you will only have to feed the kit for four weeks before it is on solid food.

Do not turn the kit on its back when feeding it (like a human baby); instead, keep the feet pointed down as it would normally feed from its mom. Turning it on its back will usually cause aspiration of the milk into the lungs. For the first few days, you will also want to gently wipe the genital area with a warm, wet cotton ball after feeding. This will stimulate defecation and urination for the baby.

This is the proper position for holding a kit while hand feeding.

REBREEDING THE DOE

So now that the doe has successfully kindled and is in the process of nursing her kits, when do you rebreed her? This is a question that is fundamental to any operation, and also one that will differ between rabbit breeders and may even vary between individual rabbits. Some facilities set a strict breed-back schedule. That is, the doe is rebred at a set time without regard to the number of kits she had or is raising. In extremely intensive farming operations, they may require her to be rebred immediately after having kits—that is, within a few hours or a few days. This is a *theoretically* possible schedule. As shown in the previous graph on the fetal weights of kits, the growing fetuses are quite small until 21 days, so they are not much of a drain on the doe while she is still nursing her current litter heavily. As her milk begins to drop off after 21 days of lactation (she reaches peak lactation 14 days after giving birth), this is just when the fetuses would be starting to grow larger.

When you add together the very light weight of rabbit fetuses up to 21 days and the fact that a doe may be receptive to a buck during pregnancy or lactation (unlike most animals), it seems as if she is built for an immediate breed-back. In reality, to rebreed this fast, you need a young doe at her prime with no more than eight kits and a feed that is very high in protein and fat. If the doe is older, or raising a larger litter, or if your feed is inadequate in any way, she will lose too much weight during lactation to sustain her through her next pregnancy. She will not be able to provide enough milk and adequate antibody production in the colostrum for her next litter. Note: anytime you are feeding high protein and high fat to a rabbit, you must balance it with free-choice hay! This is absolutely essential to prevent gastric problems, as will be seen in the next chapter.

At Chigger Ridge Rabbits, we could not find a reasonably priced feed that would sustain the doe through both heavy lactation and pregnancy simultaneously. She would invariably lose more weight than we desired if we rebred her while still lactating (except maybe at the very end). This may have been due in part to the fact that we selected for heavy milkers that could raise up to 10 or 11 kits with ease. In any case, we found it most advantageous for our operation if we let the doe have kits and nurse them through one full month before we rebred. Yes, we missed an extra litter or two a year compared to the intensive operations with our more conservative breed-back schedule, but we were able to keep does producing for as many as 4–5 years by giving them time to recover. This allowed us time to see what a doe did in the long term. We sold breeding stock as well as meat rabbits. This was not a sideline but a major part of our business plan. Thus, being able to follow a doe's performance for a year or more before offering her kits to our customers was a guarantee for us that she was a superior brood stock animal, meaning:

1. She was a reliable and easy breeder month after month and produced consistent large, fast-growing litters.
2. She milked heavy enough to sustain those litters and provide them with the all-important antibodies as newborns to give them an advantageous start.
3. She maintained her own weight, allowing for breed-back as soon as she was done lactating, proving she was able to utilize feed efficiently even under systemic stress from repeated pregnancies and lactation.

Any good, young doe from 6–16 months can do all of the above. The intensive breeding operations often only keep a doe for this period of time and then cull her as they rotate in other young does—regardless of how productive the older doe is. These are the operations that tend to breed the doe while she is still heavily lactating, as they will be replacing her soon anyway. These farms may be financially successful, as a doe will undeniably produce the best for her first year, but we did not follow their example for the following reasons:

1. These operations tend to replace nearly 100 percent of their stock on a yearly basis, and we felt that the expense and cage space necessary to raise so many replacement does was too high.
2. We were in the business of raising brood stock animals as well as meat animals and did not feel one year was adequate time to evaluate the qualities we valued.
3. Ethically, we found it difficult to constantly cull healthy, producing breeders, as there is a very limited market for older roaster rabbits.

It is true that the intensive operations avoid the problems of overgrowing nails and teeth possible in older animals. There is less chance of sore hock in younger does and more chance of damaged teats in the older. A senior doe may only produce 6–7 kits rather than the goal of 8–10. (Note: even if she has fewer eggs to ovulate as she ages, she can still be used to foster extra kits from other does.) However, all of these problems more common in an older animal can be solved though proper facility management and attention to genetic selection, as you will see in the following chapters.

WHEN TO REBREED A DOE

So should a doe be rebred a few days after the kits are weaned? A week later? Just before weaning her current kits? For us, the answer is to rebreed a doe *when she is in condition.* As mentioned, a younger doe can maintain her weight easily with a normal-size eight-kit litter. But an older doe may need a few days after her kits are weaned to regain full weight before rebreeding. If a doe raises 6 kits as opposed to 11 kits, it will obviously make a difference, as well. So, as we go to rebreed each doe, we check her body condition first. Is she back to weight? Are her teats in good shape? No other physical problems? Is she eating well and active? If all of this is positive, it is okay to rebreed. If not, it is time to evaluate whether to give her more time to recover or replace her as a breeder. We bred back between three and five weeks after having a litter, depending on the condition of the doe, which resulted in five to seven litters per doe per year.

> **We let the doe have kits and nurse them through 1 full month *before* we rebred.**

WEANING THE KITS

Kits naturally begin bouncing out of their nests a bit at about 16–17 days old. They may even begin to explore some of mom's feed at this time, but they are still almost totally reliant on milk for growth until they are a full four weeks. Some rabbit ranchers will remove the nest box at three weeks to encourage them to move around and explore other feeds. We recommend that the nest be left in until four weeks to give the kits more security. Stress from fear of new sights and sounds without a safe place to hide can actually be a systemic strain. Many rabbit raisers will wean at four weeks. Others will wean only the largest kits and allow the doe to dry off slower by continuing to nurse the smaller ones for a few days to prevent mastitis in the mammary glands (our preference). The rest will wait till the kits are older before weaning.

A doe increases their milk production from the birth of the kits until they are two weeks old, then she begins to decline in production for the next two weeks. In the wild, she can leave her nest more and more to force the kits to wean themselves as she dries off naturally. As the kits get older, they can cause serious damage to a doe's teats with aggressive nursing. In a cage, she cannot get away from them unless you remove them. Thus, when exactly to wean is a crucial decision for the kits, the doe, and the farmer. As with our breed-back schedule, we kept the needs of the animals foremost in our minds, rather than sticking to an absolutely fixed weaning schedule for all does at all times. For example, if we have already rebred the doe, it is hot, and she does not want to eat as much, and the kits do not require the extra fat from her milk to stay warm, we might recommend weaning at four weeks for the good of all. If it is extremely cold, the kits might need closer to six weeks before they are removed entirely from mom. They may not be getting a lot of milk at this time, but the added fat will help them maintain their energy requirements.

A few rabbit raisers feel a kit should be left in with the doe for a full two months or even longer and say that early weaning is "cruel." (Perhaps they have much slower-growing kits than what is found on most meat rabbit farms.) We found that many does would become lethargic and depressed and try to remain immobile to keep from attracting the kits to her if they were left in so long. She would often lose too much weight, endangering her health, and would end up with damage to her teats. We believe that keeping the kits unnaturally caged in with mom for a full two months was actually cruel to the mom.

In the wild, kits are naturally weaned as the doe leaves the nest more often in search of food. In ranching, farmers need to consider what's best for both the doe and kit.

The goal of a successful meat rabbitry: a healthy nest full of 8–10 kits every other month from each doe.

CHAPTER 4

FEEDING AND GROWING

Feed is the single most critical management decision in your rabbit operation. Feed alone is generally at least 65 percent of the cost of raising a meat rabbit, in some cases as high as 75 percent. This chapter will discuss the digestive system of the rabbit, which is absolutely necessary to understand if you are to feed correctly. (That is to say, **do not** skip this section as unimportant or boring!)

We will examine the importance of hay, the advisability of giving rabbit treats, possibilities for forage-fed rabbits, health issues caused by incorrect feeding, feed supplements, and feeding during different stages of life. In short, this chapter is the "meat" of successful rabbit raising for food. The animal must be fed correctly to produce healthy, reliable breeders and fast but safe growth in the kits, and it must be done economically in order to produce a profit for your rabbit business or an inexpensive alternative healthy food for your family.

UNIQUE FEATURES OF THE RABBIT DIGESTIVE SYSTEM

We will engage first in a little comparative physiology between animal species to highlight why a rabbit is completely different from most farm animals in digestion and why their feed composition is so critical.

Rabbits are exclusively herbivores. They dine on green stuff, as opposed to carnivores (such as a dog or cat), which eat primarily meat, or omnivores (such as humans, pigs, and rodents), which eat a bit of everything.

The carnivore, eating primarily animal protein, has the most straightforward digestive system of any animal. Their food is digested by enzymes of the gut and then absorbed in the intestine. Since they eat only meat, there is no need for large repositories of symbiotic bacteria, whose job in other animals is to help in the digestion of plant-based food. Enzymatic digestion is not efficient at breaking down plant material because of the thick cell walls found in high-cellulose plants.

Omnivores are able to eat animal proteins and enzymatically digest them like carnivores, but they can also eat and digest fats, grains, nuts, fruits, vegetables, and legumes—but *not* the high-cellulose plants on which herbivores subsist. This is why, even if a human is starving, he is unable to eat grass or tree leaves and survive for long. Like the carnivore, normal digestion of the omnivore is also primarily by enzymes in the intestines rather than through masses of symbiotic bacteria.

The herbivore has a much more complex digestive system to cope with their diet of harder-to-digest plant material. A cow, deer, sheep, or goat deals with the problem by having an enlarged portion of the stomach called a rumen, which has a large number of helpful bacteria that digest and ferment the cellulose of plants and turn it into volatile fatty acids, which are then able to be absorbed and utilized by the animal. These animals, called ruminants, will regurgitate their food (called cud) after it has been slightly digested and rechew it so that it will be more easily broken down by these bacteria.

Rabbits are exclusively herbivores.

When a ruminant eats seeds, grains, or nuts, these foods can be broken down by enzymes as they travel through the intestines, but enzymatic digestion isn't able to penetrate the thick cell walls of fibrous plants, which are the primary diet of the herbivore. Ruminants are also sometimes referred to as "foregut herbivores" since most of the digestion takes place before the food reaches the intestines. If a ruminant's symbiotic bacteria in the rumen become imbalanced by a sudden change in diet (particularly in the sugar or starch content of the feed), the animal can bloat and die.

In contrast, the rabbit, along with the horse, is referred to as a nonruminant herbivore (or sometimes called a hindgut herbivore), which means it does not possess a rumen full of helpful bacteria at the front of the digestive tract and does not chew cud. The horse is actually the most similar of the typical farm animals to the rabbit as far as digestion goes. Some of the more digestible plant material, grains, and sugars are digested enzymatically in the intestine as they pass through it, but the majority of the digestion of grass takes place in their cecum (sometimes spelled *caecum*). This is a "dead end" piece of intestine near the end of the digestive tract that is basically a large sac that allows for bacterial digestion and fermentation of the plant material, similar to the rumen in a ruminant.

All animals have a cecum, but in carnivores, omnivores, and ruminant herbivores, it is not used for bacterial digestion to any great degree and does not have a very large capacity. The cecum of a rabbit

Rabbit kits grow remarkably fast.

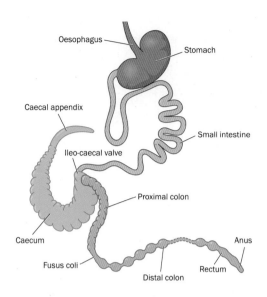

Oesophagus

Stomach

Caecal appendix

Small intestine

Ileo-caecal valve

Proximal colon

Caecum

Anus

Fusus coli

Rectum

Distal colon

Helpful digestive bacteria in their very large cecum is where the majority of the food breakdown occurs in the rabbit and other nonruminant herbivores.

is actually four times the size of the stomach and has ten times the capacity. Unlike the horse, which is naturally geared to eating small meals continually, the rabbit has a fairly large stomach for its size. This allows them, in the wild, to eat large quantities at a feeding area—often just at dawn and dusk, thus avoiding predators. Like the horse, the rabbit's cecum contains a delicate balance of bacteria in massive quantities. Since the rabbit is much smaller, has a shorter intestinal tract (the shortest of any species in relation to its size), and does not eat continuously like a horse, it has evolved a unique method of obtaining additional nutrients from its diet.

Just a few hours after eating, the rabbit feed has already passed through the intestinal tract. Some of it has been enzymatically digested in the intestines, completely undigestible parts are expelled as fecal pellets, and the remaining parts that can be digested further are diverted to the cecum to be broken down by the rabbit's cecal bacteria. At this point, the rabbit differs completely from the horse and most other mammals. The partially digested contents of the cecum are packaged into small moist pellets similar to normal fecal material but with much higher protein and lower fiber content. These pellets are called cecotropes (since they are from the cecum) and are usually expelled at night or early morning—and so are sometimes referred to as "night pellets." These cecotropes are consumed directly from the anus by the rabbit rather than deposited on the ground or cage floor. Then, these special fecal pellets pass through the entire digestive tract a second time, where proteins, nutrients, vitamins, etc., that have been released from the plant cells by bacterial digestion and fermentation in the cecum are able to be absorbed by the rabbit in the intestine. A rabbit may obtain 20 percent or more of its nutrients by this "reprocessing" of the food it eats. If imbalance occurs in a rabbit's cecal bacteria, it can result in death.

It is very unusual for either carnivores or omnivores to die from their normal diet. In contrast, an herbivore can experience difficulty even when eating its conventional diet if there is a sudden change in the type of feed, there is systemic stress, or there is an illness that interferes with the bacteria that are *vital* to an herbivore's digestion. Even the minor stress of weaning, overcrowding, nearby predators, or loud noises near the rabbitry can result in gastric problems in rabbits.

While imbalance and overgrowth of the wrong bacteria result in rumen bloat in a cow or sheep and colic in a horse, it is usually referred to as enterotoxemia or enteritis in a rabbit, and like bloat and colic, is often fatal. Keeping with one feed, using quality hay, and avoiding stressors will go a long way toward the prevention of enteritis. Kits are more susceptible to enteritis than adults. The symptoms of enteritis

A rabbit with staining around the anus has diarrhea. Enteritis is the catch-all term that means inflammation of the intestinal tract. It is often caused by change in the microflora population of the cecum from high sugar/starch diets, inadequate fiber, or rapid change in feeds.

are bloated belly, dehydration, lack of appetite, intestinal pain, and diarrhea, and if nothing is done, it leads to death. (Be aware that these are the same symptoms of coccidiosis mentioned in Chapter 2, so you need a proper diagnosis to address the correct problem.)

We should note here that the majority of antibiotics given to a rabbit for any health issue will kill off most, if not all, of their vital cecal bacteria—without which the rabbit perishes. Systemic antibiotics should never be given to your animals without consulting a veterinarian experienced with rabbit health. Ampicillin, Amoxicillin, Cephalexin, Clindamycin, Erythromycin, Lincomycin, Penicillin, Spectinomycin, and Tylosin, among others, have been shown to cause diarrhea, enteritis, and sometimes death in rabbits.

WHY HAY IS ESSENTIAL

Hay is composed of fermentable and nonfermentable plant material. That is, some parts can be digested and fermented by the cecal bacteria, and some parts can't be digested at all. With the above information on the rabbit gastrointestinal tract in mind, we will say first of all: HAY IS ESSENTIAL! No matter what it says on your pelleted feed bags about the feed being "complete" as is, virtually every rabbit expert and veterinarian will insist that hay is not an option but a requirement for a healthy rabbit. Hay provides the long fiber material that is necessary and which is not possible (in our opinion) in a pelleted feed. There are three main reasons that hay is essential:

1. Since rabbits are herbivores, constantly ripping at coarse and abrasive fibrous plants in their natural environment, their teeth never stop growing to replace the wear on them. If a rabbit is not given hay, the teeth will overgrow, and the animal will lose weight as it is unable to grind its food

properly. Eventually, a rabbit with dental problems will have to make a trip to the vet to correct it or be euthanized. In addition to hay, some rabbit raisers will give them sticks to chew to be sure they are maintaining healthy teeth.

2. A rabbit needs long fiber to prevent gastrointestinal blockage or intestinal stasis by stimulating intestinal contractions that keep the feed moving through the digestive tract and into and out of the cecum. Rabbits also groom themselves a great deal and ingest hair. Unlike a cat, which can regurgitate the "hair balls," rabbits cannot regurgitate, and without indigestible long fiber, they cannot push the hair through the intestines and expel it. This can result in intestinal blockage.

3. Fiber is critical to allow the proper bacterial health and balance in the cecum to be maintained and the natural process of cecotrophy to occur. If a rabbit is denied long fiber for a period of time, the bacteria that process it in the cecum may die off and be replaced by other bacteria that try to cope with whatever diet they are consuming.

HOW MUCH HAY TO FEED

We recommend free-choice hay feeding—that is, as much as each rabbit will consume in a day. They should be given fresh hay daily, or at least every other day. In humid weather, it can mold if left in the feeder for too long, or mice and insects may make nests in it.

Rabbit teeth grow continuously. Hay helps to keep them in good condition.

WHAT KIND OF HAY IS BEST

Timothy, Orchard Grass, Fescue, and Oat hay are all well known to be loved by rabbits and provide the good-quality, long-stem fiber that is essential. We actually found that all of our rabbits preferred a mixed meadowland hay with several grasses and even some weeds, flowers, and lespedeza thrown in. They seemed to prefer the variety of textures, lengths, and shapes. If you get a mixed grass hay, however, it still needs to be of high quality. It should have no dust or mold, should look green, and smell like fresh-cut grass.

Bermuda grass is fairly equivalent to the four types of hay listed above, but we and others have found that the majority of rabbits just do not seem to like it. Some individuals will eat anything, but we found that less than 20 percent of our rabbits would consume Bermuda in any quantity—and none if another hay was present.

Quality hay is essential to rabbit health. It should be green and smell fresh.

Many people want to use alfalfa and clover hay due to the high protein percentage— especially for growing kits and for lactating does, which need higher protein. We do not recommend either of these hays. There are three reasons we generally avoided them:

1. They crumble and produce "fines" that a rabbit will not eat.
2. These hays are too high in calcium. The calcium to phosphorous ratios are 3–4 times that of recommended hays. This high ratio may lead to urinary problems, among other health issues, especially in bucks and dry does (unbred does or does in early stages of gestation). Alfalfa is often a major component of most rabbit pelleted feeds, as it is a cheap source of high protein. However, the feed companies analyze the calcium-to-phosphorous ratio and correct the imbalance by mixing the alfalfa with higher-phosphorous, lower-calcium feeds.
3. Alfalfa and clover hays are generally much more expensive. Give your rabbits the protein they require in their pelleted feed, where there is less waste, and give them plenty of fresh, quality (cheaper) hay to munch on for their all-important fiber.

ON FEEDING TREATS

As mentioned in the preceding section on a rabbit's unique digestive system, these animals are extremely sensitive to any changes in diet that may disrupt the bacteria residing in the cecum. Giving a rabbit "treats" may lead to death unless they are introduced very slowly and are less than 10 percent of their diet. The most common types of foods known to create havoc in the rabbit's digestive tract are those that are high in starch and sugars. These create a change in the pH of the cecum, and the bacteria that process the plant cellulose die out in favor of bacteria that try to digest these unnatural foods. High-starch grains (such as corn), legumes, root vegetables, and fruits are all high in starch and sugar. Contrary to Bugs Bunny munching carrots all day, carrots are actually rarely eaten in the wild.

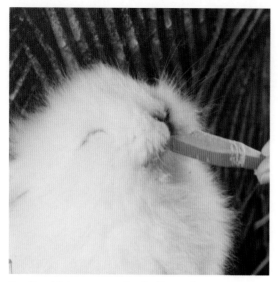
Feeding high sugar or starchy "treats" to young rabbits may actually lead to death.

If you are going to give your rabbit supplemental treats, give them green leafy vegetables that are low in calories and high in fiber, such as spinach, carrot tops, kale, strawberry leaves, parsley, radish tops, cucumber leaves, dandelion leaves, beet greens, or watercress, to name a few. With proper drying, other garden produce can be stored for future feed treats, such as corn stalks, pea or bean vines, and sunflower stalks and heads. You should not leave fresh greens in the cage for longer than about an hour. Harmful bacteria love to multiply on wilting greens that have been stomped on by dirty rabbit paws. For this reason, such foods are hard to feed in a large, commercial-type rabbitry. These plants are high in water, and none are high in the protein needed by the growing rabbit or lactating doe, so there will not really be a big difference economically to your operation if you are feeding garden scraps or not.

Unless you want to do an in-depth study on animal nutrition and try to create a plant diet that is high enough in protein and calories, has the all-important fiber, contains no plant toxins, and has all the proper ratios of vitamins and minerals, we recommend that supplemental fresh food treats be kept at a minimum.

Throwing greens onto the cage floor to be trampled and contaminated will lead to a sick rabbit.

ARE "FREE-RANGE" RABBITS POSSIBLE?

Trying to raise your rabbits on grass alone is not really an option, and we hope to explain why below. Some people use "rabbit tractors" similar to chicken tractors (see Chapter 5). These are basically movable cages set on the ground, where the rabbit can munch out on grass and weeds. The cages must be movable, as a rabbit will de-nude an area very quickly. If it is not moved daily or every other day, it soon becomes unsanitary, parasite-ridden, and a health threat to your rabbits.

So you might ask, "If rabbits eat grass in the wild, why can't I just raise all my rabbits in a rabbit tractor system and not spend any money on feed?" There are two reasons at least why this is not possible:

1. A rabbit in the wild does not eat *only* grass. It is free to range around and eat bark, various weeds, flowers, and even bushes to balance its diet. A rabbit that is caged can't do this.

2. A rabbit in the wild will usually only produce one or two litters in its life before it is killed by some predator. A domestic rabbit that is being used to raise food for humans will be asked to produce many litters every year and thus must be given the proper diet to stay healthy and happy as it does so. An all-grass diet just is not sufficient for a doe bred several times a year.

We were often asked about growing free-range rabbits in "well-fenced" fields. We have never seen anyone do this successfully in a pasture-type setting. There are simply way too many rabbit predators. Even if you have fences to keep out coyotes, dogs, foxes, cats, etc. (assuming they can't dig under the fence), there are still flying predators (i.e., eagles, hawks, and owls) and ones that can climb fences (opossums and raccoons) or go right through fences (weasels and snakes). Every single one of these critters loves munching on baby rabbits, and many will take full-grown rabbits as well.

So is there any way at all to fill the marketing niche of rabbit meat that is raised on "natural," not pelleted, feed? The only exception we have seen may be in the use of sweet potato forage. Dr. Lukefahr of Texas A&M University's rabbit research program conducted a study of rabbits fed sweet potato forage or sweet potato hay versus commercial rabbit pellets. Sweet potato greens are very high in protein

Aerial predators as well as four-footed ones threaten "free-range" rabbits.

compared to most other greens and are high in fiber. Thus, unlike most greens, they can actually be utilized as a diet for rabbits in place of pellets and hay. The one problem is that, like all greens, they are also very high in water. The rabbits on the sweet potato greens still had to be fed 1.2 ounces of oats to meet energy requirements and given a mineral lick for trace minerals and salt. Dr. Lukefahr's research showed that the sweet potato forage-fed rabbits were only slightly lighter at harvest compared to the pellet-fed rabbits (3.2 pounds carcass weight compared to 3.5 pounds when fed the pellet feed). However, the profit margin for the sweet potato-raised rabbits was nearly double that of the pellet-fed rabbits due to the greatly decreased feed costs.

For those rabbit raisers who have established a niche market for sustainable agriculture or forage-fed organic meats and are having trouble meeting protein requirements for economical production with their rabbits, the sweet potato might be your answer. We suggest carefully reading the research of Dr. Lukefahr for additional information (Lukefahr et al. 2012).

EVALUATING YOUR COMMERCIAL RABBIT PELLETS

Commercial feed companies have trained animal nutritionists on their staff whose job is to create a food that is palatable for the rabbit and that meets all of their dietary needs during all stages of growth. Rabbit pellets do not generally contain any animal proteins, antibiotics, or hormones objectionable to many consumers.

Below is a picture of a typical rabbit feed tag you may find at your local feed store or cooperative. You should be able to obtain a basic feed such as this for between $12–$18 for a 50-pound bag, depending on where you live. Many feed stores and farmer's cooperatives will give a discount for orders in bulk, although in the summer in humid areas you must watch for mold in stored feed. If you can't find a basic feed for a maximum of $18 within a reasonable driving distance, you will need to investigate exactly how much your market will be willing to pay for your rabbits to be sure you can cover costs and make a profit (see Chapter 10).

NET WT. 50 LBS. (22.67 KG.)

50X6

18% RABBIT PELLET

For grower and breeder rabbits.

CAUTION: Use Only As Directed

GUARANTEED ANALYSIS

Crude protein (Min)	18.0000%
Crude fat (Min)	1.5000%
Crude fiber (Min)	16.0000%
Crude fiber (Max)	18.0000%
Calcium (Ca) (Min)	.8000%
Calcium (Ca) (Max)	1.3000%
Phosphorus (P) (Min)	.4500%
Salt (NaCl) (Min)	.5000%
Salt (NaCl) (Max)	1.0000%
Vitamin A (Min)	2000.00iu/lb

INGREDIENTS:

Forage products, processed grain by-products, roughage products, plant protein products, grain products, molasses products, calcium carbonate, salt, DL-methionine, ferrous oxide, ferrous sulfate, vitamin E supplement, choline chloride, calcium pantothenate, riboflavin supplement, niacin supplement, vitamin B-12 supplement, vitamin A supplement, cobalt carbonate, vitamin D3 supplement, manganese sulfate, ethylenediamine dihydriodide, zinc sulfate, copper chloride, mineral oil, sodium selenite.

RUMINANT MEAT AND BONE MEAL FREE
22CE G 50X6

Notice on this example tag that the ingredients are very generic—forage products, plant products, grain by-products, etc.—rather than specific feed items. This lets the feed mill vary the ingredients based on what is cheapest for them to obtain at the moment. Since it is often a problem for young rabbits to have sudden diet changes, changes at the mill may cause unexpected trouble for the rabbit farmer. Feed labels with exact ingredients listed are thus preferable.

This is 18 percent rabbit pellet—meaning that it is 18 percent protein. Protein is necessary for growth, milk production, and reproduction. Late gestation and lactation require a minimum of 18% protein if you are breeding the doe on a schedule to maximize production. Kit growth also requires high protein: 16–18 percent. Your bucks and dry does would actually do better on a

A pet rabbit might do fine on hay and more "natural" foods, but a working doe that produces several litters a year needs a diet formulated to meet her increased requirements. Commercial feed companies have trained animal nutritionists whose job is to create such feed.

lower-protein feed (13–15 percent); however, it is often uneconomical to maintain several different feeds in your operation. If you want to keep to just one feed type, you can feed slightly less of a high-protein feed to your bucks and dry does and give them lots of quality hay.

Fiber is the next thing you want to look at on the feed tag after protein. Fiber is essential for a healthy gut. Even though you are feeding hay, we found that our rabbits had fewer GI problems with a feed that was a minimum of 16 percent fiber. Although feed tags will not give you this information, most of the fiber in pellets is more likely to be fermentable fiber, which will be digested or fermented in the cecum by the rabbit's normal bacteria. A feed too low in this fermentable fiber may lead to an imbalance of bacteria in the cecum that must be avoided to help prevent enteritis. Remember, this fiber does not replace hay, which is essential for their dental health and to maintain intestinal movement, preventing GI stasis or blockage.

Fiber in a pelleted feed does not replace hay.

Vitamin A is listed under the guaranteed analysis, and vitamins D and E are listed under ingredients of this example feed tag. These are three essential vitamins that a rabbit must have in their food (vitamins K and B can be synthesized by the microflora in the gut.) Hay, stored for any length of time, begins to lose its vitamin content. A deficiency of these vitamins may result in impaired reproduction, retarded growth, birth defects, and weakened immune systems. Whenever you have a problem in your rabbitry, make sure to take your feed tag when you talk with your veterinarian.

The next thing to look at in your feed is the percentage of fat. Fat is an important energy source (especially for the lactating doe). It aids in the absorption of the fat-soluble vitamins (A, D, E, and K) and promotes a healthy, glossy coat. This feed label example shows that there is 1.5 percent fat in this feed, which is on the low side. Both baby and adult rabbits are very efficient at utilizing fats for energy. That is why rabbit milk has one of the highest fat percentages of any animal milk and why a doe only needs to feed her young once or twice a day. However, it is often hard to find economical rabbit feeds that are high in fat. Researchers have found that the ratio of pounds of feed per pounds of weight gain

Calf-Manna is a great lactation supplement for the doe and protein supplement for kits.

Black oil sunflower seeds are another popular supplement for adding extra fat and vitamin E to the ration. Vitamin E helps to keep the immune system strong and protect against bacterial and viral infections.

(feed conversion) decreases with increasing fat (Arrington and Franke. 1974), resulting in significant feed cost savings. Yet it is hard to find a rabbit pellet with even a minimum of 3 percent fat, and 5 percent would be even better for lactating does. So what do you do if you only have low-fat feeds available for a reasonable price in your area? There are two options that we explored in our operation: Manna Pro Calf-Manna and black oil sunflower seeds, discussed next.

FEED SUPPLEMENTS TO CONSIDER

The addition of Manna Pro Calf-Manna (www.mannapro.com) to the diet of lactating does and kits beginning to eat solid food was used in our operation with great success. That's right, Calf-Manna, which, though formulated for calves, is actually labeled for many species, including rabbits. A lactating doe only needs a tablespoon of this feed supplement daily, and a growing kit only a teaspoon. Calf-Manna provides added fat to help supplement the inadequate levels in many commercial rabbit pellets. It is considered a very "energy dense" supplement. A doe with a large litter that is milking heavily is often simply physically unable to eat enough regular commercial pellets to meet all of her energy needs. Calf-Manna also contains 25 percent very digestible crude protein. The combination of increased energy and high protein helps to increase milk production without a decrease in the animal's condition.

The example commercial feed tag on page 68 is also rather low in vitamin A for a lactating doe—only 2,000 IU per pound . Experts vary in their recommendations for this important vitamin, but you will often see everything from 2–3 times this recommended for lactating does. Calf-Manna contains 20,000 IU per pound of vitamin A and will help increase this essential vitamin in your rabbit ration. There are excellent rabbit feeds out there that are made specifically for the lactating doe, but these are often far too expensive for a commercial meat rabbit operation. We found that a lower-quality but affordable commercial pellet, supplemented during heavy lactation with the Calf-Manna, was an economical alternative.

Do not overfeed Calf-Manna, however, on the assumption that "more is better." Because of the high vitamin content, feeding more than the recommended amount for rabbits can lead to vitamin toxicity. Curiously, vitamin A deficiency and toxicity symptoms are identical—poor conception rates, small litters, stillborn kits, kits born weak with slow growth rates, and sometimes hydrocephalus (swollen head) at birth.

Calf-Manna contains anise, a very aromatic spice with a licorice-like flavor. We often referred to Manna as "rabbit candy." We never found a single animal that didn't like it! It is great for getting animals back on feed or introducing kits to solid food, as well as being a supplement for lactating does. We would always offer our rabbits their Calf-Manna first—never mixed in with their regular commercial pellets. Many individuals like the Manna so much that they will scratch out all of the other pellets from the feeder, looking for the "yummies." We would even offer animals that didn't need the supplement, such as our bucks and dry does, just a few pellets as a treat each day as a simple way to check on their health. If a rabbit won't run up to the feeder to eat their Calf-Manna, chances are it is very ill!

A second supplement we used in our operation to increase our productivity was black oil sunflower seeds. These are the black seeds you can find as a wild bird feed. The black seeds, not the striped ones, are higher in the nutrients needed by your rabbits. They are becoming a popular supplement for horses, cattle, swine, goats, and chickens as well. They are (depending on the source) 25 percent fat, 15 percent protein, and 40 percent fiber. They are also very rich in vitamin E, which is essential for the immune system. We found them particularly attractive in the winter months, as the high fat content was such a good source of energy to help the animals combat the cold, rather than increasing overall rations. For this same reason, we tended to use them less in the summer months to keep the animals from getting too obese. This is also one supplement that is relatively easy to grow yourself and decrease your costs. The seeds, once dried, can be stored for feeding in the winter, while the leaves, stems, and dried heads can be given to the rabbits for a nice high-fiber treat.

Though not quite as universal as the Calf-Manna supplement, nearly all of our rabbits loved their sunflower seeds, and like the Manna, we had to feed them separately rather than mixed into their regular ration to prevent them from wasting the commercial pellets as they looked for the sunflower seeds to munch. You do not want to go overboard with this supplement either; 8–10 seeds for a dry doe or buck is more than enough, or you can get them too fat. The heavily lactating doe and fast-growing kits can utilize between a teaspoon to a tablespoon, however.

> The lactating doe's feed is one of the highest costs you will have in your operation.

HOW MUCH TO FEED

A Californian or New Zealand buck, dry doe, and pregnant doe are all fed similarly. For most on 16–18 percent commercial pellets, this means 4–6 ounces (by weight) of feed daily. Whether they are on the lower or higher end of this range depends on the individual. The breed, body size, activity levels, and the time of year all impact feed utilization. It is important to monitor each animal individually to keep them

Only use creep feeders for larger kits; smaller ones will crawl inside and contaminate the feed.

at optimal weight. This means weighing them, or at the very least, picking them up and feeling their body condition on a routine basis. The ribs, spine, and pelvic bones will feel sharp in an emaciated rabbit, and the rump will look curved in. These bones will be easy to find in a rabbit of ideal weight, but they will have rounded-feeling edges. In an obese rabbit, it is hard to feel the rib and spine bones at all, and the rump will look rounded or even bulging. If you overfeed an individual, it will have difficulty breeding—and you are also wasting money on your feed bill. At the same time, you can't starve a profit out of any animal! Our motto is always: "You can't manage what you don't measure," so keep those scales handy.

The lactating doe's feed is one of the highest costs you will have in your operation. Incorrect feeding will affect her crop of kits, the length of time until she is able to be rebred, her overall health, and your profit margin. Lactating rabbits, as a rule, should be fed free-choice hay and pellets—that is, as much as they can eat. We found that if we increased the feed slowly over the first three days of lactation, we had a much lower incidence of mastitis (inflammation and infection of the breast).

The doe will be at the height of lactation in only two weeks, and she will need all she can eat at that time. However, for the first couple of days, the newborn kits are often simply not large enough to drink all she might produce if full feed and feed supplements are pushed immediately after kindling. The result of forcing full milk production before the kits are large enough to consume it all can result in milk stasis in the breast—a known cause of mastitis. Since the doe has eight teats, if she has less than eight kits, she is even more susceptible if feed is pushed too soon.

Since the doe is usually energy-starved immediately after kindling, we immediately double her feed the first day (from 4 to 8 ounces). Then, at day two, she will get 10 ounces. At day three, she is fed 12 ounces and begins a teaspoon of lactation supplements (Calf-Manna and black oil sunflower seeds). At day four, we give her *all she will eat* of her 18 percent commercial pellets and continue to increase the supplements up to a tablespoon of each over the next days. A good New Zealand White that is a heavy milker will usually limit herself at 16 ounces (1 pound) of feed a day at the height of lactation, but we have had some eat as much as 24 ounces (1.5 pounds) with a large litter.

The feed should always be replaced with fresh feed every 24 hours (rabbits will not eat damp pellets, and moldy feed can kill), so attention must be paid each day to how much she is consuming to prevent waste. This becomes a very important consideration in a larger rabbitry. If you have 50 milking does, for each doe *not* eating 3 ounces of feed a day, that is 50 x 3 = 150 ounces per day wasted. Multiply that by 30 days in a month, and you have 150 x 30 = 4,500 ounces per month wasted. Since there are 16 ounces in a pound, that means 4,500 ÷ 16 = 281.25 pounds of feed thrown away! That is well over five 50-pound bags of feed a month! No operation can absorb that kind of continual loss.

Toward the end of lactation, it would be ideal to back the doe off of full feed to allow her to dry off slowly to prevent mastitis, while still maintaining access to free-choice pellets for the kits. There are rabbit "creep feeders" available, which are basically long feeders that have several holes too small to allow the doe to get her mouth in while the kits can. We tried these feeders but found them to be unsanitary, as the kits love to squeeze into the holes and would defecate and urinate in the feed. Rodents and insects were also attracted to the dark of these feeders.

Another problem we encountered with the creep feeders was that, although the doe could not eat the kit's feed, the kits could still get to hers. They would often mob and empty her feeder before moving on to theirs—thus starving mom. We found that a staggered weaning schedule seemed to be a better answer if we wanted the doe to dry off slowly. We would remove the largest of the kits first and allow a couple of the smallest a few days of extra milk.

We found a staggered weaning schedule is the best way to prevent mastitis in the doe. It also gives slower-growing kits a few days with less competition to catch up to the weights of their littermates.

FINISHING THE FRYER

Growing kit rabbits are always given all the pellets and hay they will eat in a day. As seen in the following graph, the feed needed to grow a kit rabbit (green line) begins to increase rapidly at about 4–6 weeks and is one reason this is considered an optimal weaning age for commercial meat rabbits (in addition to allowing rapid breed-back of the doe).

If a normal weaning weight is 1.5 pounds and the target weight of the finished fryer rabbit is 4.75–5.75 pounds, the rabbit needs to more than triple its weight after weaning. It needs to do this before 16 weeks of age to be considered a fryer animal by most processing plants. Also note on the graph that those last few ounces of weight gain as the rabbit gets older requires far more feed (green line) for each ounce of rabbit weight (orange line) than when a rabbit is younger. If you are able to move your operation from a 12–14 week harvest of a fryer to a 10–12 week harvest, the savings in feed costs are

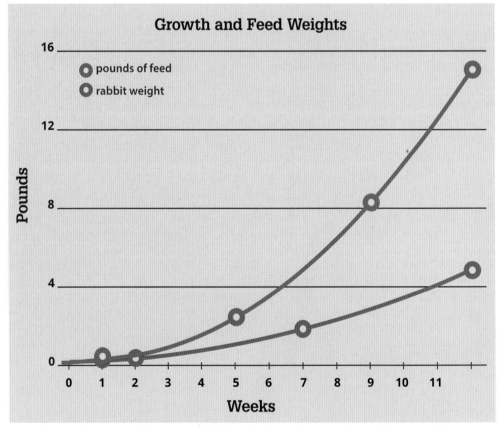

It can be seen that at 8 weeks it takes a LOT less feed to add a pound of weight to a rabbit compared to at 12 weeks. The younger the rabbit reaches marketable weight, the more profitable the operation.

very significant—so significant, in fact, that this one point may make the single difference between a profitable and unprofitable operation.

The graph presented is from a non-intensive meat rabbit farm harvesting 5-pound rabbits from 8–10 kit litters at 12 weeks. The *optimal* goal is actually a 5-pound meat rabbit in 8 weeks. However, to (realistically) achieve this ideal target requires 8 kits or less in a litter from a prime young superior doe, optimal (higher cost) feed, and climate-controlled barns.

PRACTICAL FACTORS CONTRIBUTING TO A FASTER FINISH

The first consideration is the actual finishing weight one aims for. As will be mentioned in Chapter 9, this will depend on your final market. Many live rabbit buyers for meat packing plants require between 4.75 and 5.75 pounds. Those buyers who will only accept 5.75–6.25 pounds (live weight) will often lose their rabbit growers due to the ever-increasing feed needed to hit this higher weight.

If you are selling dressed rabbit meat direct to the public as we often did, you must find the point that is good for that market as well. A dressed rabbit carcass weight is generally about 50 percent of the live weight if sold as a whole carcass. (Ours were higher than that, but 50 percent is an easy way to estimate things mathwise.) Thus, a 6-pound live-weight rabbit will dress out at around 3 pounds. We found that many people who have never eaten rabbit may be hesitant to buy a larger rabbit when sold by the pound, yet they might try out a smaller, cheaper one. But in our opinion, anything below 2.5 pounds carcass weight simply does not have a good enough meat-to-bone ratio and will not build a satisfied market. Therefore, for our direct sales, we shot for a finishing weight of between 5 and 6 pounds (2.5–3 pounds carcass weight).

Our objective, as a profit-seeking meat farm, was to hit a finishing weight of 5–6 pounds in as short a time as possible while utilizing feed that is reasonably priced. Besides the use of feed supplements mentioned above, one fairly simple strategy to achieve fast growth is the use of terminal sires, such as an Altex or Flemish Giant mentioned in Chapter 2. These larger sires, known for their fast early growth, can give you a week to ten days earlier finish time.

A second tactic is measuring the performance of individual animals and selecting superior breeding stock. For example, using litter weaning weight measurements helps to select for heavier milking does. This is considered a very heritable trait. An even more vital consideration, however, is the measure of the kits' rate of growth once they are weaned and on solid food only. Since all the kits are on free-choice, identical feed, comparing this growth can help you to select lines that produce the fastest-growing kits. For more in-depth strategies, refer to Selecting for Economically Important Traits in Chapter 6.

In summary, raising meat rabbits as a viable economical food source for your family, or to sell, requires constant monitoring of your operation. Compared to most other meat farming, rabbit raising has a very fast turnover of animals, in which you can raise several generations in a single year. Pregnancy is short, lactation is brief, and the time to harvest is less than four months from the breeding date. The results of the selection of superior stock and small changes in farm procedures can be seen in mere months rather than years. It has a downside in that, if you do not have the time to spend on the necessary daily surveillance and tracking of your animals, the result can be enough waste in feed costs to turn a profitable and rewarding enterprise into an unproductive and frustrating one.

CHAPTER 5

CAGES AND HOUSING

CAGE REQUIREMENTS

The USDA Animal Welfare Regulations (Part 3, Subpart C, Section 3.53; see reference section) suggest that a meat-rabbit–size female (8.8–11.9 pounds) with a litter should have 6 square feet of floor space. Therefore, if your cage is 2 feet deep, it should be 3 feet long (2 x 3 = 6). A buck in this same weight range, according to this same source, can get along with as little as 4 square feet. However, we found that they tended to maintain a better weight by giving them a bit more room to move around—so 6 square feet is good for them as well. Please note that if you are raising rabbits for pets or research rather than meat, you will be regulated under the US *Animal Welfare Act,* which *requires* specific cage sizes—and they will inspect you! (See Chapter 8 for more information.)

A breeding rabbit doe needs 6 square feet of space in a cage that protects her from the elements, predators, and disease.

When first weaned, the kits can all be moved as a single litter of 8–10 kits into a single 6-foot-square cage. However, to prevent overcrowding they should be separated into two such pens as they grow, with no more than five rabbits per cage for finishing. After four months of age, the females you are keeping for breeding stock can be housed with 2–3 per pen, but the males will begin to fight and must have their own cages. We suggest that the cage be at least 18 inches high for New Zealands or Californians to allow the rabbit to sit up or hop around with ease (taller for the giant breeds).

Our preference is to buy or build identical cages of 2 feet by 3 feet for all of our rabbits. Alternatively, some people will make larger group cages for their "grow-out" rabbits that are being finished for meat. If you are planning on this strategy, you should include multiple food and water sources so there is not too much competition.

BUILDING CAGES

The first thing to consider when planning for your rabbit housing is how many cages you will need—always more than expected. For a minimal sustainable operation—that is, to provide rabbit protein for an average family without buying replacement breeders constantly—four does and two bucks from unrelated lines are required. However, this does not mean simply six cages. For this six-rabbit scenario, you will need at least thirteen 2-foot by 3-foot rabbit enclosures if you are alternating the breeding of two of the four does at a time—six cages for the adults and four for the growing meat rabbits. Then you will want to add a quarantine cage and a pen or two for your future replacement breeders.

For a 100-kit-per-month operation, you will need to breed around 15 does per month. Yes, 15 does times 8 kits per litter is 120 kits, not 100, but you must take into account that one or two may not conceive, may have kits out of the nest, or may have a small litter. Remember also that the doe is pregnant for a month and nursing for a month, so that means 30 does are actually needed to breed 15 of them each month (if you follow our recommended breed-back schedule from Chapter 3).

Overcrowding can lead to stress and disease.

You also want to breed these does at roughly the same time to prevent having some kits too large and others too small at harvest. If you are using a buck only one time per day, you need 5 bucks to be able to breed all 15 does in 3–4 days. This brings the total to 35 cages for the adult rabbits in a 100-kit-per-month operation. You should also have around 25 cages for your 100–120 growing meat kits if you count 4–5 kits per pen. This brings the total to 60 (6-foot-square) cages for this size operation. Again, you will need extra cages for isolation, replacement animals for yourself, and for holding breeding stock for sale (if that is part of your marketing plan).

A surprising number of people would tell us that they were just planning to "raise the kits in with their mother" and thus did not need all those additional cages for them. This is not recommended for several reasons:

To prevent disease, fryers must be kept in sanitary conditions and with no more than five growing animals in a cage that is 6 square feet.

1. Although the *goal* is to hit your target harvesting weight at exactly eight weeks; this often does not happen. If the mother is rebred when the litter is at four weeks and she is ready to kindle, but her former litter still needs another week or two before they hit marketable weight, what do you do without pens to move them to when the mother is ready to drop the next litter?

2. There is also the minimal requirement of cage floor space (discussed above). Unless you have *very* large cages for your does, there is definitely not adequate room for a dam and 8–10 kits as they approach harvest weight.

3. Some does have no difficulty weaning the kits themselves as they hit 4–5 weeks. Others cannot wean on their own and will allow the kits to continue nursing as long as they are in the cage, even if they are almost as big as her! This can lead to damaged teats and an out-of-condition doe. Often, the moms will become depressed, go off feed, and sit in a corner, trying not to move so as not to attract the "hungry hoard."

Once you have determined the number of cages you will need for the size operation you plan, it is time to build or purchase them. Cage wire will usually be the main part of your cage. The sides should be strong enough to resist any local predators (i.e., strong enough to withstand opossum, raccoon, and dog teeth). For the side and roof, 16-gauge 1-inch by 2-inch welded wire is usually adequate. Look for wire that has been galvanized *after* welding to help prevent rust problems. Chicken wire is not strong enough to resist determined predators and will rust easily when exposed to rabbit urine.

We will also mention here the type of wire referred to as "baby-saver wire." This is wire intended for the sides of cages. The bottom four inches have a smaller-size mesh (white arrows in photo on page 80). The purpose of this is to prevent newborn kits from falling out of the cage if they end up out of the nest. The kits can crawl surprisingly long distances, even when first born, if the mother kindled outside of the nest box. Occasionally, a kit will also fix so strongly onto a teat when feeding that it is actually dragged out of the nest when the mother leaves it—the greedy little bugger! Both of these scenarios are fairly rare, and one must weigh the extra cost of the more expensive baby-saver wire against the occasional loss of a kit.

Cages using "baby-saver" wire with smaller-sized openings at the bottom prevent newborn kits from falling out of the cage.

The floor of the cage must hold up to 40 pounds for a 6-foot-square cage to be used by 3–4 young does or one doe with a nursing litter—this means stronger, 14-gauge galvanized wire. Usually, 1 inch by ½ inch is used on the floor for the comfort of the rabbit's feet, while still allowing the manure to drop through the wire.

Vinyl-coated wire is an option when considering cage material. It is more expensive but more resistant to rust from urine. However, it is not amenable to burning off hair during cleaning as wire alone is (see the following section).

HOUSING CONSIDERATIONS

Housing for rabbit cages can be varied and still be quite successful, so we will not exhort one specific kind. Instead, we will include here the minimum requirements we feel are important for a successful, cost-effective operation, whether it is for commercial farming or for private rabbit meat consumption. This means housing that:

- Protects the animals from the elements and predators
- Is constructed to maintain sanitation and good health, while taking into consideration the animal's physical comfort and social needs
- Is built to minimize work for the farmer
- Is economical in the long term

PROTECTION FROM THE WEATHER

The first requirement of any housing scheme for any livestock is protection from the elements. For rabbits, this means relief from heat and protection from cold winter winds. A rabbit is far more adversely affected by hot rather than cold temperatures. Over 95°F to a rabbit can be fatal (though rarely), especially if they are stressed by other factors.

The simplest way to minimize heat is to locate your rabbits in a shaded area. This reduces solar radiation on the rabbit and keeps them cooler. So, if possible, locate your animals in a well-ventilated barn or shed, under trees, or on the shady side of buildings. If you have no natural shade and no building, you will need to provide heat relief in the cage itself by setting roofs high enough to allow heat to escape but still provide protection from direct sunlight.

Running electricity to your rabbit housing to operate fans or other cooling equipment is optimal. Other tricks for heat reduction during severe heat waves include lightly misting them with a water bottle, using a misting hose set in front of a fan, or giving them a frozen water bottle in the cage to lie next to. You must balance heat reduction with the rabbit's preference for low humidity, however. Also, if you are misting during the day, you will have to feed at night to prevent wet feed, which rabbits will not eat.

It doesn't have to be fancy or expensive. This small, two-sided shed has wire to prevent predators and is located under trees for shade.

Protection from cold is less important for survival of adult rabbits than protection from heat. Healthy adult rabbits can survive temperatures below 0°F if protected from winds and kept dry. There is less loss of kits if they are in an area that is protected from cold drafts. Protection from the cold will also reduce winter feed requirements, as you do not have to provide as many extra calories for generating body warmth.

While protection from cold winds is important, it must be done in such a way that ammonia levels do not build up from a rabbit's strong urine in enclosed spaces. This can lead to respiratory problems and is known to be directly connected to the clinical incidence of a highly contagious disease known as snuffles (as mentioned in Chapter 2). You need to maintain a balance between adequate shelter and adequate ventilation. One economical idea we used successfully is layering burlap over chicken wire as wind barriers. This can be used to cover barn windows or large barn doors, or placed on an exposed side of a freestanding hutch—providing wind protection, yet still allowing ventilation.

If you have a pole barn, with a roof but no walls at all, an inexpensive option for winter protection can be to stretch heavy, clear plastic across the top of the building (where walls would be) and burlap over chicken wire across the bottom—in effect, creating low-cost temporary walls in winter. This allows sunlight to be concentrated by the plastic to provide additional warmth and yet still allow good

Protection from cold is less important for survival of adult rabbits than protection from heat.

"Walls" of burlap over wire can furnish winter wind protection while still providing essential ventilation.

ventilation at the bottom where ammonia gas (which is heavier than air) settles. In the summers, you just remove the top plastic to keep them cooler.

Another viable alternative to fully enclosed barn walls is using roll-up blinds or shades that can be raised or lowered depending on the weather and provide either shade in summer or wind protection in winter.

If you do have your rabbits in a tightly enclosed barn or shed that doesn't have a lot of ventilation, you will need to set up fans on each end. One is to draw air in, and the other is to expel air out. The inward air fan should be located higher in the wall than the exhaust fan. You want fresh air to flow down over the rabbits and then toward the floor and out, never the opposite way, drawing the ammonia-laden air up from the floor to the rabbits. In a tightly enclosed barn, these fans may need to be in operation both summer and winter.

If you find a doe constantly remaining in a nest box with her kits in very cold temperatures, it is not because she is a good mother and trying to keep them warm; it is that she is too cold herself. She may end up with damaged teats by constant nursing in these circumstances. She should be provided with a second box to warm up in. Alternatively, set some bedding outside of the nest or provide a heat lamp if you have electricity available. DO NOT locate a heat lamp directly over a nest box with kits, however, as this can make the nest too hot for the kits and too hot for the doe to want to nurse them. Locate the lamps instead to shine in one corner of the cage, thus allowing the doe to warm up or not as she chooses. Of course, it goes without saying that all electric wires must be rigged in such a way that the animals cannot reach them to chew them, and the lamp itself set so it won't risk catching anything on fire.

PROTECTION FROM PREDATORS

Once you have a good location for your rabbits, protected from excessive heat and winter winds, the next thing you need to ensure is protection from a rabbit's natural predators. Since nearly everything will prey on a rabbit, this can be a daunting requirement. A typical 1-inch by 2-inch welded-wire cage, enclosed in a freestanding hutch or located in a barn, will protect against most predators. The snake is one exception. A common rat snake (found almost everywhere in the United States) can easily climb and will fit through 1-inch by 2-inch wire and raid a doe's nest of newborn kits. Fortunately, a snake of this type will only eat about a kit a week or even every two weeks, and with the rapid growth of kits, if it tries to eat too large a kit, it may find itself so bloated as to be unable to get back through this size wire to digest its prey. Such snakes must be removed and relocated to prevent recurrent predation—even though they are great at rodent control.

Snakes have no problem eating eggs or rabbit kits whole.

The panic response of a rabbit must also be taken into account when protecting against predators. If animals such as cats, opossums, or raccoons prowl around the cage at night or climb up on top of the cage in an effort to get to a rabbit (even if it can't), this may trigger the flight response of the rabbit, resulting in the rabbit attempting to run in a panic from the danger. Such a panic response can result in a broken neck or back. Enclosing your rabbit cages inside of electric fencing or using guard dogs may help deter such predators. Motion detectors set to turn on spotlights or radios where predators enter may also help. Of course, trapping and removal of persistent varmints may be needed.

The floor of a rabbit pen is typically 1-inch by ½-inch welded wire. A bit of rabbit toe can hang down below the cage floor. It is enough that an enterprising opossum, raccoon, dog, or fox can nip off toes if the rabbit has nowhere to get off of the wire floor to safety. Some of these predators will actually eat dozens of toes in a single night! (Yes, we have unfortunately seen this . . .) To prevent this, you should provide a solid surface as well as wire floor (such as a resting board, as noted at the end of this chapter) unless you have a completely predator-proof perimeter.

For those who have ideas of farming rabbits in open areas, you will encounter constant predator threats to your animals, including aerial predators. Birds of prey are difficult to guard against, and they are protected by federal law, making it impossible to trap and remove them or shoot them. We do not, therefore, recommend any "open-air" rabbit raising. Even if given structures to hide under from aerial predators, you are still essentially confining an animal in a situation where it cannot leave a threatening environment. You are forcing your animals to become furry "sitting ducks." Although they may look happy hopping around in a "free" situation rather than in a secure cage, it is not humane!

MAINTAINING SANITATION

The third requirement for housing in a rabbit operation after protecting them from the elements and from their natural predators is having enclosures built to promote sanitation and prevent disease. This means, first and foremost, having a cage design and materials that can be easily and adequately disinfected. Wood, with its porous nature, is impossible to disinfect properly. No, even soaking in bleach is not adequate! In our opinion, no wood should be used to construct rabbit hutches, cages, nest boxes, or feeders. Not only is it impossible to disinfect, but rabbits will chew wood, and thus the longevity of cages built of it is questionable. Wood will become saturated with rabbit urine and lead to odors that may get your rabbitry shut down (by angry neighbors) in all but a totally rural setting. Metal (galvanized to protect from the strong urine), tile material, and hard plastics are the materials of choice for the construction of rabbit cages and hutches to provide surfaces that can be adequately disinfected.

Another consideration when planning your cages is easy access to all areas, especially the corners, which most rabbits will use as a toilet area. Even though theoretically the rabbit manure is supposed to fall through the wire floor, not all of it will. A larger dropping may get stuck, or a rabbit may step on the droppings before they fall through. If there is excess hair from rabbits preparing their nests, this may clog up some of the wire holes. Whatever the cause, there will be a need to clean the rabbit cage floor—especially the toilet corner—on a regular basis. We recommend that pens not be over two feet wide and have a door located so that you can reach all areas of the cage for easy disinfecting.

> **No wood should be used to construct rabbit hutches, cages, nest boxes, or feeders.**

MANURE HANDLING

Your plans for manure disposal will have a huge impact on the housing systems and designs you choose. A larger operation can become extremely labor-intensive if this factor is not well considered during the initial startup. Removal of rabbit "gold nuggets" for use, disposal, or preparation for sale should be well thought out before you buy your first rabbit. We will discuss below the design considerations and pros and cons of the most common housing options with regard to manure disposal, as well as animal health and practical farming considerations.

MOBILE RABBIT TRACTOR

The green cage in the photo on page 85 depicts one kind of rabbit tractor (or Morant hutch). This is the name given to any movable rabbit cage located on the ground that is shifted every day or two and the manure then cleaned up or left to fertilize the fields. There are a million different designs of rabbit tractors that can be found on the Internet. They can be made of welded wire with PVC pipes, other plastic, or metal supports and should have lightweight roofing. The main requirements are that it provides shade and protection for the feed, protects the animal from the elements, and is on wheels or light enough to be easily moved to prevent buildup of manure in the cage. This photo is an "Eglu Go Rabbit Hutch" from Omlet (www.omlet.us) and is the best design we have seen to date.

The pros of the rabbit tractor are that the cages are separate from the other rabbits, thus reducing the possible transmission of disease from rabbit to rabbit. It provides a natural surface environment that prevents foot sores such as sore hock and furnishes access to natural browsing. The cons are that water must be supplied in bottles or bowls that are subject to winter freezing and bacterial contamination and take a lot

This is one of the better "rabbit tractor" designs— protecting from the elements and predators. It is sanitary if cleaned inside and moved regularly.

of time for cleaning and refilling, as opposed to automatic watering systems. Because of frozen water scenarios, we suggest this type of system for summers only in the colder climes. In warmer regions, this could be used year round.

The tractors absolutely *must* be moved on a regular basis, and the water checked daily, which adds time to a farmer's workday. There is the potential for rabbits digging out or predators digging into these cages unless the bottom of the run also has wire, such as in this Eglu design (this also eliminates the does digging holes to try to nest in the outside run).

The main concern we have in using them is that there is a possibility of disease or parasite introduction from being on ground that might have been previously used by wild rabbits, cats, chickens, dogs, rodents, etc. When the rabbit industry moved from ground rearing to cage rearing up off the ground, the incidence of disease fell dramatically.

This particular rabbit tractor is easily cleaned and provides very good protection from the elements, while still having excellent ventilation. Lifting the tractor enough to move it may result in young kits getting injured or escaping. Older rabbits can learn to move inside the lockable box for moving, but kits are likely going to bounce around all over the place! It may be used in a small family farm situation, but it is not recommended for larger enterprises.

FREESTANDING RABBIT HUTCH

The typical freestanding rabbit hutch, such as depicted on page 86 (left panel), is a very common housing system. It will sometimes have catch pans that are removed and cleaned, but usually, the manure is removed manually from under the hutch for composting or use on gardens. Alternatively, worms can be raised under the hutch for manure disposal.

This enclosure, like the rabbit tractor, has the advantage that the rabbits can be kept more separated than in a barn, and thus prevent the spread of any diseases. This particular hutch has good protection from wind and provides adequate shade and feed protection. However, it is made of wood, which cannot be disinfected, may lead to odors, and which rabbits will chew.

Like the rabbit tractor, this cage system cannot be set up for automatic watering easily or economically. The hutch shown has limited access to natural light, which may contribute to reduced productivity during short day-length seasons. Both rabbit tractor and freestanding hutch systems are more common for the smaller family farm rather than a larger commercial operation, due to the time

The old-style wooden rabbit hutch does provide adequate shelter and protection from predators, but it is not the most sanitary or efficient housing system.

A more modern adaptation of the old wooden freestanding hutch can be used for commercial rabbit ranching.

and problems of ensuring constant fresh water. That being said, there are more modern versions of the rabbit hutch, as seen in the above photo (right), that have been adapted for commercial use.

CAGE BANKS

Cage banks inside barns are the most common housing system for any larger or commercial rabbitry. The advantage of this system is that electricity can be run for cooling equipment and lighting for day-length regulation (important for breeding, as seen in Chapter 3). Electricity also means that heated automatic watering systems can be set up to provide clean, fresh water year round in any climate with very little work on the farmer's part. Barns or sheds offer shade and protection from wind and predators. The disadvantages of a barn with cage banks is that if a disease is introduced, it can be spread easily from cage to cage. Ventilation must be adequate to combat ammonia from numerous rabbits in a confined space.

The top photo on page 87 shows a single-tier bank of cages set up in an old two-story tobacco barn and rigged over worm beds. There are electric lights, an automatic watering system, and large barn doors to provide breezeways in summer. Old tobacco barns are an excellent choice to be "repurposed" for rabbit raising. The strong crossbeams that once held tobacco can be used for suspension wires to provide cage support, and the tall building construction allows heat to rise up and away from the animals.

Worms are a classic way to compost rabbit manure and provide additional income at the same time. A complete guide to worm farming is beyond the scope of this book; however, we will mention some of the pros and cons of using worms for manure disposal. The main advantage is the extra income generated and the sustainability of being able to use the worm-converted rabbit manure as organic soil enricher. The con to this system is that you can overly contaminate your worm beds with disinfectant and might feel limited in the amount of cleaning you can do in your rabbit cages.

Providing cover for the worms while cleaning helps. A dry wire brush removes any manure stuck on the cage floor. A handheld propane torch can be used to burn off excess fur on the cages and reduce the cleaning needed, but you should never skimp on disinfecting just to keep your worms happy! If worms are not doing well under the pens, you may have to remove the manure to a separate area and grow your worms there.

Cage banks can also be set up for a hose-out system for manure disposal. The double-tier outfit pictured here was purchased from a commercial supplier (assembly required). We used a 4-inch PVC pipe to connect to the drains under the cages and run it outside the shed for manure disposal. A double-tier system such as this allows for raising a large number of rabbits in a relatively small barn or shed. A hose-out system is ideal for providing an ammonia-free environment and easy disinfection of pens.

Caution must be taken to adjust your ventilation to prevent too much humidity in the barn while washing out, as this can lead to respiratory ailments and moldy food. In a double-tier system such as the one pictured, the bottom cages are lower to the floor, and cleaning *must* occur on a regular basis to prevent the buildup of ammonia (which is heavier than air) in these cages. Care should also be taken that the upper bank of rabbits does not contaminate the feeders in the lower bank. Urine guards (the metal strips seen at the bottom of the upper rack of cages in the photo) help defend against this. A few of the more obstinate rabbits may still manage to urinate or defecate into the bottom feeders—almost as if they are aiming for them! Metal feeder covers should be kept on hand for use on the bottom tier feeders if needed.

Cage banks over worm beds eliminate work and provide additional income.

Ammonia from urine is heavier than air and will sink to the floor. When using two-tier caging, being able to wash out completely and regularly is essential to animal health in the lower cages.

We also found that manure and urine can freeze overnight in winter in facilities that are not climate-controlled. This means hosing out during the warmer parts of the day if there could be an overnight freeze. If there are several days of subfreezing temperatures, this problem is compounded. Although there will be no ammonia during the time when the waste is frozen, there will be an immediate need for a thorough (and substantial) cleaning during thaws. We thus recommend this type of system be used only in a climate-controlled facility in colder areas. These double-tier cages can make a climate-controlled facility more economical, however, since double the number of rabbits can be raised in the same space to help pay for the heating/cooling costs of the barn or shed. We have seen triple-tier used as well, but if you are a short rabbit raiser, this will increase your time and effort with lots of going up and down on step stools.

KEEPING THE NEIGHBORS HAPPY

Rural farmers can set up their rabbits with just the rabbits' needs in mind, but the more suburban backyard breeder or mini-rancher must also keep the neighbors in mind when planning the rabbit facility. Since rabbits are quiet, the main concern of neighbors will be any odors or pests. Making sure that your manure is drained into gravel-based beds and kept dry will keep odors down.

If your cages are located over hard clay, you may have to dig out a layer of clay and set up drainage gravel and sand substrates. Sprinkling the manure with lime will also help with flies and odors. As mentioned, no wood should be used near a rabbit cage, as it will absorb odors. Since rabbit urine can spray out a bit, this means that cages should not be too close to any wood walls (or the walls should be covered with metal, plastic, or tile). Care must be taken to keep feed in metal containers, and rodent control measures should also be taken.

If there is no odor, no vermin, and no sound, some neighbors may still be upset just seeing caged animals—especially if they know they will be used for meat. If possible, locate your rabbits in an area of your property where neighbors will have to trespass to see them. If there isn't enough room to do even that, it might be worth the investment to put up a latticework wall around your rabbits if they are not in an enclosed building. Happy neighbors can neither see, hear, nor smell your animals!

DOMESTIC RABBIT SOCIAL NEEDS

Housing seems to be the best place to make a note about rabbit social requirements. The kit is definitely a social animal. It does best in groups, and when possible, should be moved in litter groups to any new cages. They are naturally nervous creatures, as well as playful, and enjoy the safety and security of *their* gang. As they get close to four months, however, the "teenage" males will become aggressive to other males and want to breed every female in sight. They need their own cage at this point, for the peace of their fellows. The females, however, can stay in groups of 2–4 until near breeding age, or a young doe can stay in with her dam if the dam is not bred to kindle again immediately. Once a female is at breeding age, she should be given her own space, and she will get very territorial as she makes it hers.

All rabbits are naturally timid, or they would not survive in a world designed to eat them; they thus prefer routine and their "safe space" over human definitions of freedom. A domestic rabbit has not been "free" for hundreds of years and is usually terrified if presented with freedom. They no longer have adaptations for survival in the wild.

Kits love to put dirty paws into water bowls—a sure way to promote disease.

WATER SYSTEMS
WATER BOWLS

There are basically three different watering systems for your rabbits. The cheapest is simply a water bowl. We do *not* recommend water bowls for meat rabbit rearing—period. The animals will contaminate them without fail. Young rabbits, in particular, will inevitably put their paws in them after walking through manure. It is the quickest and most infallible method of promoting disease! There is also the necessity of constant washing and the problem of frozen water in winter unless your facility is climate controlled. If the dish or bowl is spilled on a very hot day, you could actually lose rabbits from dehydration. If you absolutely have no choice but to use a water dish, there are ones available that can be attached to the cage side and up a bit off the floor to keep spillage and contamination to a minimum.

WATER BOTTLES

A better choice is using rabbit water bottles. This avoids the problem of contamination, but there is still the drawback of the time spent in washing and refilling them and the possibility of freezing and bursting in winter. Heated bottles are expensive and rarely last long. If the bottles are located in the sun, they tend to grow algae, making cleaning even more vital. One bottle will not be enough for a nursing doe and kits for a day. You must use two bottles or purchase special adapters for two-liter soda bottles.

Water bottles are better than a water bowl but are really only practical in a smaller-size rabbitry due to the time involved in cleaning and filling. Water bottles tend to leak fairly often and will keep the manure under the cages wet unless catch pans for the liquid are added and emptied regularly. The use of both water bowls and bottles adds to the time that will be spent in the rabbitry. This should be factored into your setup plans if you are planning on more than a few rabbits.

Water bottles are not practical for raising large numbers of rabbits, as they are prone to freezing, algae, and take significant time to clean and refill.

AUTOMATIC WATERING SYSTEMS

Automatic watering systems are ideal for prevention of disease and reduction of work for the rabbit grower. They can be as simple as a covered five-gallon bucket of water, hand-filled every few days, in which gravity feeds the water through pipes to the lick valves. By hooking this bucket up to a pressurized water source, such as found in homes, and adding a float valve (similar to those found in the back of a toilet tank) to enable the water to refill automatically, you guarantee that the animals will never be without fresh, clean water even if you are away from the farm (see photos on page 91).

Automatic watering systems are ideal in climate-controlled facilities or in regions that never see a freeze, but in below-freezing temperatures in facilities that are not climate-controlled, they can be a challenge. There are two basic arrangements that can prevent freeze-ups. One is adding a small heating element to the main water vessel with a circulation system, similar to a boat bilge pump, so the warm water is continuously pumped through the pipes to the cages and back to the warm water reservoir. The reason that a pump must be used is that if the temperature drops very low and the rabbits don't happen to be drinking enough to pull the warm water through the pipes, you can get freezing of the pipes even if there is hot water in the main water source. Any pipes without warm water inside must be insulated well.

There are temperature-controlled devices called thermo-cubes that can be purchased to plug in your water heating systems. They are cheap and programmed to come on automatically when the ambient temperature drops below 37°F. You simply plug the thermo-cube into the outlet, and the heater and

A heating unit in the main water receptacle and a recirculating pump prevents frozen water lines.

recirculating pumps into the cube, then you don't have to worry about the weatherman being wrong or about remembering to turn on the systems.

However, we found (the hard way) that for rabbit cage banks with longer than around 40 feet of water piping, even these recirculating systems may not be adequate in very cold (single-digit) temperatures. The distance is simply too great from the heat source in the main water reservoir to the end of the water line to prevent freezing in the pipes—even when the water is kept moving with a recirculation pump.

A system we utilized for long banks of cages, rather than the recirculation arrangement, was running heat-wire(s) through the pipes themselves (with some of the wire coiled up in the main water vessel to keep it warm as well). This system never froze up on us, even at zero and below. The only downside you might encounter is that, with the wire running *through* the piping (PVC), if it ever

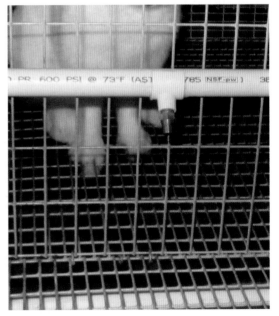

Either gravity or a circulation pump feeds water through pipes to the lick valves.

Any pipes without warmed water inside should be insulated for winter.

An alternative to recirculating pumps is heat wires in the piping.

> **Make sure that all animals have access to fresh, clean water *at all times*.**

needs to be replaced, the entire water system and piping would have to be dismantled. We never encountered this problem. But just to be sure, we ran two wires, one for backup, through the pipes. Only one wire is used unless it gets *really* cold. Two wires will make the water hot enough for tea. (In the photo above, you can see these wires before we put the piping together.) This heated wire system can be used with PVC pipe, but not in the Edstrom or Borax flexible watering tubes that some people use instead of PVC. The flexible tubes, however, are reported to be more resistant to busting in freezing situations.

Even with the pipes and main water source warm, the lick valves themselves may still freeze up. Most times, the rabbits will warm them enough to get water by repeated licking. Alternatively, they can be thawed by simply warming them with your hand or a hair dryer.

As you can see, rabbit water, no matter what the source or time of year, must be checked regularly to make sure that all animals have access to fresh, clean water *at all times*. In the end, it comes down to the type of cages, the initial investment you are willing to make, and the temperatures expected in your particular part of the country to determine which watering system is best.

We strongly suggest contacting the experts at either Bass Equipment Company (www.bassequipment.com) or Klubertanz Equipment Company (www.klubertanz.com) for this important decision. Both companies have been in the business for over 40 years, and their experts are always willing to discuss your specific needs and sell you the best system to meet those needs.

We always found them willing to help with answering questions. (They also sell the heavy-duty galvanized welded wire for building your cages or whole cage systems, as well as nest boxes, feeders, and almost anything else needed in a rabbit operation.) There are other reputable rabbit equipment dealers, of course, but these are the two we dealt with almost exclusively for our operation and have no hesitation in recommending.

OTHER EQUIPMENT

While we are discussing equipment, we will briefly mention feeders. There is no way to completely eliminate the possibility of feed contamination by soiled rabbit paws. Kits are notorious for sticking their feet in the feeders or climbing entirely in them if they can. Freestanding food bowls in the cage are more subject to being fouled or spilled than feeders that are attached to the cage wall in some fashion.

For larger facilities, the ease of feeding also becomes an important time consideration. Feeders that are attached to and filled from the *outside* of the cage are optimal to save all the time of opening and closing dozens of pens each day. We recommend metal or hard plastic feeders with screen bottoms to allow the "fines" from pelleted feed (small particles from broken pellets) to drop outside the cage through the bottom and reduce the need for cleaning the feeders. Without a means for the fines to drop through, these particles will build up (rabbits will not eat them). This can attract insects and rodents, and can lead to toxic mold if they get moist.

> **It is important to meet all of your animal's psychological, as well as physical, needs.**

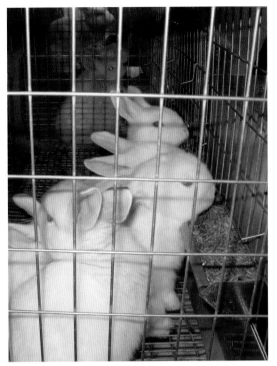

Pellet feeders should be located to allow easy feeding, limit waste, and avoid contamination.

If not prevented, kits will use feeders as bathrooms. Large open pans are not recommended.

Always use hay feeders to provide this all-important feed in a sanitary manner. Locating both hay and pellet feeders outside the cage saves time in feeding.

Hay can be thrown on the top of the cage for adult rabbits, but young kits are too short to reach the top and will need a hay feeder. Hay should *not* be placed inside the cage on the floor where it can be

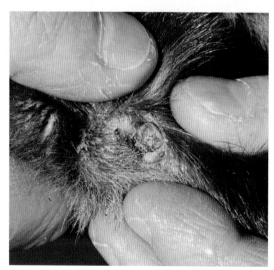

Though wire floors are optimal for sanitation, some heavy meat rabbits require a resting board to help prevent ulcerations on the feet known as sore hock.

soiled. Again, hay feeders that are situated on the outside of the cage aide in easier, faster feeding for larger operations. Both hay and pellet feeders are available from Bass Equipment Company or Klubertanz.

The addition of "resting boards" are also recommended for rabbit operations. This is somewhat of a misnomer, as it is not a wooden board. (We do not recommend wood for *anything* due to the inability to properly disinfect it.) These mats should be big enough so that the rabbit can get all four feet on them to allow relief from wire surfaces and help prevent foot sores known as sore hock (see Chapter 2). Such relief is recommended from a humane farming point of view, even if there are few incidences of sore hock with your stock. The resting board has to be regularly cleaned, but the comfort of the animal must

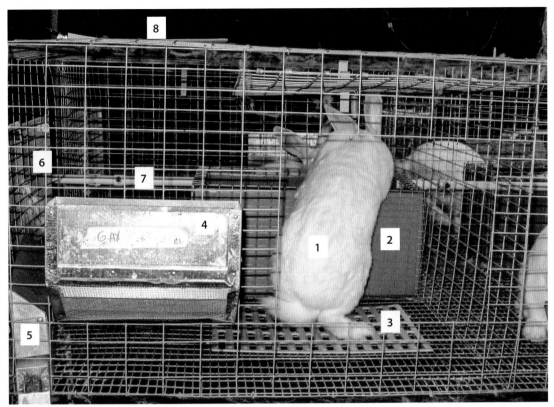

1) Healthy rabbit; 2) nest box (Universal Sani-Nest); 3) resting board; 4) feeder that is filled from outside cage with screen bottom; 5) kit creep feeder (not generally recommended); 6) hay feeder that is filled from outside the cage; 7) lick valve in automatic watering pipe; 8) clipboard for barn notes on top of cage.

take precedence over the time needed for cleaning. We used simple 12-inch by 12-inch ceramic floor tiles, which can be purchased from your local building supply store. These can be very easily scraped clean with a snow scraper, spatula, or cement trowel. A quick wipe with a disinfecting cloth finishes the cleaning. There are also plastic boards sold by the rabbit equipment companies that have drainage holes to help keep them clean.

Finally, we will mention rabbit toys. Yes, toys! It is important to meet all of your animal's psychological and physical needs. Not every rabbit will play with toys, but some desire the added stimulus (especially young rabbits first moved to their solitary adult cages). You do not have to spend a lot of money to provide for this need. Sticks, empty toilet tissue or paper towel rolls, empty soda cans or plastic water bottles all make fine toys. We would often shop yard sales for hard-plastic baby rattles, as these are always made with safe paints and intriguing shapes.

Above is a photo of a New Zealand White meat rabbit in a typical commercial cage bank housing setup, with everything she requires.

CHAPTER 6

SELECTION AND GENETICS

INBREEDING, LINE BREEDING, OUTCROSSING, AND CROSSBREEDING

It is important to fully understand these four terms and the consequences of using (or avoiding) them in your rabbitry to safely fix desired traits in your rabbits. The percent of co-ancestry (or genes shared) between related individuals is shown in the table on page 98.

INBREEDING

The term "inbreeding" refers to the practice of repeatedly breeding the most closely related individual animals together: mother/son, father/daughter, or full siblings (brother/sister). Close inbreeding is evaded by nearly every mammal on earth by physical, social, or psychological avoidance mechanisms. For example, packs of lions, wolves, bands of horses, and primates all chase off young males from the group and thus prevent them from breeding with closely related female relatives. A person who practices deliberate inbreeding on their farm believes that by limiting the possible genetic variation among their rabbits, they will get to their goal of certain selected traits faster. This is actually true, so why is inbreeding so bad?

Percent of Co-ancestry (or Genes Shared) between Related Individuals

RELATIONSHIP	PERCENT RELATED
Parent/Offspring	50%
Full Siblings	50%
Half Siblings	25%
Grandparent/Grandchild	25%
Great-grandparent/Great-grandchild	12.5%
Aunt/Uncle/Nephew/Niece	25%
First Cousins	12.5%
Half First Cousins	6.25%
Second Cousins	3.13%
Third Cousins	0.78%

Inbreeding = breeding rabbits that share 50% co-ancestry.
Line breeding = using rabbits with 12.5% co-ancestry.
Outcrossing = breeding two individuals with <1% co-ancestry (but same breed).
Crossbreeding = breeding together two different breeds with 0% co-ancestry.

There is, unfortunately, a phenomenon known as "inbreeding depression" caused by genetic mutations that leads to diminishing fitness of an animal to survive. It is caused because the normal shuffling (and elimination) of unfavorable genes cannot occur with the reduction of overall genetic material when closely related individuals are bred together. Inbreeding depression has a wide variety of physical effects, including:

- Smaller adult size
- Slower growth rates
- Loss of immune system function
- Higher newborn death
- Reduction in litter size
- Decrease in sperm viability and female fertility
- Elevation of genetic abnormalities (one of which will be discussed in more detail later)

LINE BREEDING

Line breeding is a mild form of inbreeding that all breeders of purebred animals (purebred = animals of a particular breed; i.e., the New Zealand White) practice to some degree. This is because all *purebred* stock derives from a few individual foundation animals. However, a person practicing line breeding as opposed to inbreeding will make an effort to breed more distantly related stock. A person practicing inbreeding will breed animals that share 50 percent of their genetic makeup, whereas a line breeder breeds animals that share only 12.5 percent of their genetic material (such as first cousins or a great-grandfather to a great-granddaughter).

Breeding together rabbits that share 25 percent of their genetic material is considered inbreeding by some and line breeding by others. Line breeding is commonly practiced in rabbit facilities where a very few superior bucks or does may be used to "fix" certain genetic traits that the farmer is trying to establish—hopefully, without the introduction of harmful genes or the effects of inbreeding depression.

OUTCROSSING

Outcrossing is the breeding of two animals of the same breed that have no common ancestors within 5–6 generations. A person practicing outcrossing will make an effort to breed only unrelated or very distantly related stock (below 1 percent shared genetic material).

On occasion, many farmers will outcross with other rabbit breeders that have unrelated purebred stock and are selecting for the same traits. Outcrossing is done to avoid accidentally creating carrier animals of some hidden abnormal gene and to keep from slipping into inbreeding. You can also create outcrossing opportunities on your own farm by buying your initial stock from several unrelated farms and then keeping them separate genetically for several years as you select the best stock from each group. Later you can begin breeding these groups together to outcross.

CROSSBREEDING

Crossbreeding is the term used for the breeding together of two totally different breeds, such as a Golden Retriever dog to a Poodle, an Angus bull to a Hereford cow, or a Flemish Giant rabbit to a New Zealand White. When crossbreeding to a separate breed, you widen the pool of unique genetic material and generally get the opposite of inbreeding depression: hybrid vigor. This usually results in more rapid growth and an increase in immune system function or general robustness.

If using a completely inferior animal during crossbreeding, hybrid vigor effects will not necessarily overcome the poor performance of that parent. Breeding superior animals of two different breeds helps to counter the inbreeding and line breeding effects inherent in any purebred domestic stock. However, when you go overboard and start to breed a bunch of crossbreeds together (creating a "mutt"), you begin to move into an area of unpredictability for the offspring since your genetic pool has become *too* wide, resulting in such things as staggered growth rates and variations in carcass quality.

AVOIDING GENETIC ABNORMALITIES

One fairly common example of a visible genetic abnormality caused by too much inbreeding in the New Zealand White rabbit is the expression of a recessive "Bu" gene (from the word "buphthalmic," meaning large eyeball). This abnormality results in bulging eyes, sometimes with a whitish film and early onset

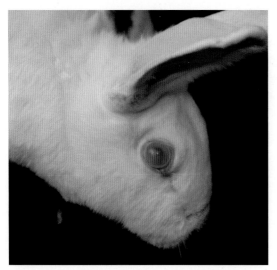

One genetic abnormality found in the New Zealand White is an inherited early onset glaucoma from a common recessive gene scientists named the "Bu" gene, for Buphthalmic (meaning large eyeball). Inbreeding with an unknown carrier of this gene can result in a large number of blind rabbits surprisingly fast.

glaucoma that results in blindness. All animals have two copies of any gene—one from the mother and one from the father. Only animals with *two* abnormal Bu genes will actually develop glaucoma. Those kits with only one abnormal copy of the gene will not develop glaucoma themselves but are called "carriers" of the abnormal gene and might pass this recessive gene to their offspring. You cannot tell they are a carrier just by looking at them.

If two carrier rabbits are bred together, *some* of the offspring will end up with two recessive copies—one from the dam and one from the sire. These offspring will suffer from this inherited glaucoma. Both the buck and doe lines that produce such kits (or any genetic abnormality) should be culled as breeding stock even if they don't show any signs of the abnormality themselves. The reason for this can be explained by using a genetic prediction tool called a Punnett

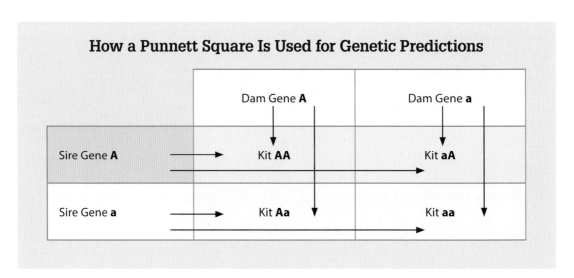

How a Punnett Square Is Used for Genetic Predictions

	Dam Gene **A**	Dam Gene **a**
Sire Gene **A**	Kit **AA**	Kit **aA**
Sire Gene **a**	Kit **Aa**	Kit **aa**

This illustration is of a genetic tool called a Punnett square, used to predict the genetic makeup of offspring for a particular gene variant. Here, we are looking at a dominant gene (denoted as **A**) and a recessive version of the same gene (denoted as **a**). In this example, both parents have one copy of each variant (the dam and sire are both genetically **Aa**). Follow the arrows to each kit's genetic result. Each kit actually represents 25 percent of a litter. This tool can be used to predict such diverse characteristics as a genetic defect or coat color.

square to help visualize the percentage of animals that may be affected or might be carriers of such a gene if left in the breeding population.

In the Punnett square table below, the capital B stands for the dominant or normal form of the Bu gene, and the lowercase b stands for the recessive or abnormal form of the Bu gene (causing glaucoma). To use the Punnett square for prediction, simply take the gene (B or b) from the dam column above the kit and add the B or b from the sire row to the left of the kit to predict the kit's genetic result. Remember, the dam and sire both *appear* completely normal in this case, but each *carries* one copy of this abnormal gene.

When two carrier rabbits are bred together:

- Kit #1 will have two normal (BB) copies of the Bu gene and will not pass on the abnormal gene to any offspring (that is, 25 percent of the litter will be completely normal).
- Two kits in four, or 50 percent of the litter (kits #2 and #3 in the Punnett square), will not *have* glaucoma but will *carry* the abnormal gene for it (genetically they are Bb).
- One kit in four (kit #4) will actually develop glaucoma since they only have abnormal copies of the gene (bb).

Kit results from breeding two carrier animals together	Dam's dominant or normal **B**	Dam's recessive or abnormal **b**
Sire's dominant or normal **B**	Kit 1 with **BB**: both copies of gene are normal—no glaucoma	Kit 2 with **Bb**: one copy of gene is normal and one abnormal—no glaucoma, but a carrier
Sire's recessive or abnormal **b**	Kit 3 with **Bb**: one copy of gene is normal and one abnormal—no glaucoma, but a carrier	Kit 4 with **bb**: both gene copies are abnormal—glaucoma

When both parents are carriers of this kind of genetic defect, 25 percent of their offspring will be normal, 25 percent will have glaucoma, and 50 percent will look normal but be carriers of the defect.

Note: these are only probable percentages; it will not be *exactly* this in every litter. (Just like if you flip a coin four times, you will not always get two heads and two tails despite the 50 percent probability for each outcome.)

In a second example, if instead of breeding two carrier (Bb) parents together as shown in the first Punnett square, only the buck you are using is an unknown carrier of the abnormal gene. The doe is not a carrier but is completely normal (BB).

These results are different in that you will not even realize that you are breeding this defect into your operation, since there will be *no* animals born that actually develop glaucoma—but 50 percent of the offspring will actually be carriers! When breeding these offspring together, in a few months, you will

Kit results from breeding one carrier animal and one normal animal	Dam's dominant or normal **B**	Dam's dominant or normal **B**
Sire's dominant or normal **B**	Kit 1 with **BB**: both copies of gene are normal—no glaucoma	Kit 2 with **BB**: both copies of gene are normal—no glaucoma
Sire's recessive or abnormal **b**	Kit 3 with **Bb**: one copy of gene is normal and one abnormal—no glaucoma, but a carrier	Kit 4 with **Bb**: one copy of gene is normal and one abnormal—no glaucoma, but a carrier

When breeding an animal that is physically and genetically normal (in this case the doe) with an animal that looks normal but actually carries a genetic defect (the buck), ALL of the offspring will appear normal, but in reality half of them will carry the genetic defect and can pass it on in the future.

suddenly have 25 percent of your rabbits developing early onset glaucoma—all because you were using a buck that looked normal but was a carrier of this abnormality.

Glaucoma is only one illustration of a problem gene known in meat rabbits; others include:

- Cataracts
- Hereditary incisor teeth misalignment
- Polycystic kidney disease
- Lymphosarcoma susceptibility
- Lysozyme deficiency
- Neuroaxonal dystrophy

In summary, you want to avoid inbreeding of individuals that are very closely related, while selecting for the desired trait you are trying to improve in your stock by using line breeding or outcrossing (with other purebred breeders selecting for the same trait). Crossbreeding between two different breeds results in hybrid vigor in the first generation with increased robustness and growth, but continuous multiple crossbreedings may introduce a less predictable product.

SELECTING FOR ECONOMICALLY IMPORTANT TRAITS

We have commented throughout this book on the importance of the selection of superior replacement breeding stock. Some of the factors to consider for this stock are:

- A strong immune system, as shown by keeping stock that shows no evidence of clinical diseases, even if exposed to carriers of disease
- Strong maternal instincts in the females, as evidenced by nesting ability
- Good milking ability in does, as measured by heavy kit-weaning weight
- Fast-growing kits, as determined by tabulating the time to finish a fryer or the weight gained from weaning to finishing

- Good feed efficiency, demonstrated by determining the actual feed needed by kits to gain a pound of body weight, and keeping the kits that need less
- Physical characteristics necessary for good carcass quality

Note that anything you are selecting for has to be able to be measured in some way, and the trait has to be heritable. For example, according to some well-known and highly respected rabbit researchers (McNitt et al. 2013), conception rate and litter size (which *many* rabbitries try to select for) are actually much less heritable than milk production and kit growth rate.

Unfortunately, it is almost impossible to fix these crucial qualities in your rabbitry by trying to select for all of them at once. It is important to prioritize and select for just one at a time. This is because, unlike inherited glaucoma controlled by the single Bu gene, the heritability of commercially important traits such as kit growth rate, milking ability, or efficient feed utilization are considered to be "polygenetically determined"—any one of these three traits may be governed by multiple genes. If each trait was theoretically controlled by 10 different genes, selecting for just one trait at a time—for example, fast kit growth—would require the breeding animal selected to have as many of the 10 genes favorable to fast kit growth as possible. Trying to select for three such traits at once would require the breeding animal to have 30 favorable genes, which becomes nearly impossible to achieve.

You should look at your operation and determine what is affecting your productivity the *most* and begin there. Are your feed costs too high because it takes too long to move the meat animals from weaning to acceptable finishing weights? Do you have trouble with high litter loss from females that do not make proper nests? Do you have young females that have difficulty milking heavy enough to raise a full eight-kit litter so that all kits hit a good weight at weaning? Do you always seem to be battling some disease or ailment in your facility? Do you have to feed more than you think you should to keep your adults at a good weight?

Since there are no genetic tests available to tell you exactly what these economically important genes are, the farmer has to depend on recording data on their animals—in regard to the trait they are wanting to improve—and retain breeding stock that demonstrate these genes based on the data that is recorded. Once you have selected for your first genetic trait and are happy with your results, you move on to selecting for the next trait you want to enhance in your stock, and so forth. If you are just starting out, you should closely question the breeder from whom you are considering purchasing your foundation stock to learn what they select for and why.

Assuming that you have obtained good, healthy commercial stock that are nesting and raising eight-kit litters to harvest (or you have already selected to reach this first goal), it is time to select animals to help your farm's movement toward cost efficiency. Remember, feed is your single greatest cost in raising rabbits. There are four measurements you can use to begin the selection of stock to reduce your feed costs.

1. Overall farm feed conversion ratio
2. Doe milk production
3. Kit feed efficiency
4. Kit rate of growth

A scale is *the* single most important management tool in any rabbitry. Remember, you can't manage what you don't measure!

OVERALL FARM FEED CONVERSION RATIO

Overall farm feed conversion ratio is the number of pounds of feed it takes to make one pound of rabbit on your farm. If you refer back to the chart shown in Chapter 1 on feed efficiency of a rabbit versus other species of animals, you will see that the rabbit's feed efficiency is a range of 2–4:1. This means it takes between 2 to 4 pounds of feed to make 1 pound of rabbit. It is very helpful for you to know just where in this range your farm falls. The measurements take a bit of work, but we recommend that you calculate this number at least once every year to make sure you are moving in the right direction.

Begin by weighing the feed you give your doe rabbits every single day from the moment the kits are born until they are weaned. You also need to weigh the uneaten feed that you throw out every day to subtract that out to have the actual feed consumed. When the kits are weaned, you continue to measure all the feed they are given (again with all the waste subtracted out). At harvest, you weigh the kits. To calculate your feed conversion ratio, divide the number of pounds of feed eaten from birth to slaughter by the pounds of rabbit kits raised.

For example: you fed 50 kits (and their moms while nursing them) a total of 980 pounds of feed from the time they were born until harvested. You weighed the kits at harvest and have a total of 263 pounds of live rabbit. Divide 980 by 263 to get a feed conversion ratio of 3.7 for your farm (980 ÷ 263 = 3.7). It takes 3.7 pounds of feed at your farm to grow 1 pound of rabbit.

Note that this calculation is to obtain data for your farm only. It is solely to determine how your rabbits are doing as time passes. You can not claim your stock is genetically superior because this measurement is better than John Smith's farm, 500 miles away, which uses totally different husbandry practices.

Obviously, the fewer pounds of feed, the better. If this number is going up from year to year and not staying the same or hopefully going down, you need to look at your operation a bit more carefully. Determine if you have a parasite or other medical problem on the farm, whether or not you are keeping too many unproductive senior does, whether you are inbreeding too much, or if you have changed feed to a brand that is not adequate. If none of this appears to be the case, collecting more specific data, as described below, may help you rectify the situation.

DOE MILK PRODUCTION

Measuring your doe's milk production is done by measuring the growth of the kits from birth to weaning. This does not involve any feed measurements. Simply weigh the entire litter of kits at 24 hours old and then again at 28–30 days old (or whenever they are weaned). Subtract their birth weight from their weaning weight, and you have a direct measure of the doe rabbit's milk production (by the pounds of kit grown on her milk). This is the time period when almost all of the growth is from nursing and not any other feed source. This number will help you to determine which does (or which lines) are the best milk producers, and since this is a heritable trait, it can be used as a selection criteria.

However, this is where careful and complete records are essential if you are to determine which lines are best to use for your future brood stock. Comparing the milk production from a four-year-old doe to a nine-month-old doe in her prime will not tell you which genetic line is actually better in terms of lactation. A four-year-old will never produce as much milk as a prime nine-month-old doe. Instead, look at the weaning weights of the litter of the four-year-old back when she was a young doe of nine months to get an accurate comparison between the two genetic lines.

You also need to take into consideration the time of year for barns that are not climate-controlled. A midsummer litter in the warmer climates, when the doe is too hot to eat as much, or a midwinter litter in cold climes, when a portion of the doe's feed must go to energy requirements against the cold, can lead to lower weaning weights in general than fall or spring litters, when the temperature is more moderate. For appropriate selection, you need to compare lines that have measurements taken at the same time of year. Of course, the litter sizes should be identical as well; you can use fostering to match them up.

KIT FEED EFFICIENCY

The kit feed efficiency and rate of growth are two traits that will financially impact you more than any other, and that you can positively affect by selection of breeding stock lines that display good numbers. The kit feed efficiency is very similar to what was calculated above for the overall farm feed conversion ratio but takes the mom out of the picture. You will weigh the kits at weaning and begin to measure all of the feed consumed by the kits at that time. Then weigh the kits at harvest. Subtract the kits' weaning weight from their harvest weight to get just the pounds of kit that has been grown from weaning to harvest.

For example, if a kit weighs 1.5 pounds at weaning and 5.5 pounds at harvest, you have 4 pounds of kit growth. To obtain the kit's feed efficiency or feed conversion, you divide the total pounds of feed consumed by the kit from weaning to harvest by the 4 pounds of kit produced in this time period. Then select for lines that need less feed to make a pound of live meat.

It is not necessary (or practical) to house each kit separately for these calculations. For example, if a litter weighs a total of 12 pounds at weaning and 45 pounds at harvest, the litter gained 33 pounds during that time period (45 - 12 = 33). If the kits consumed 125 pounds of feed during that time, that litter's feed efficiency is around 3.79 pounds of feed per pound of body weight gained (125 ÷ 33 = 3.79). Draw your future breeders from the litters that have the lowest number.

KIT RATE OF GROWTH

If you do not have the time to measure daily feed intake, a simpler measurement is the kit rate of growth. Remember from Chapter 4, that as a kit gets older, it needs more feed to produce a pound of meat. To calculate rate of growth, you weigh the kits at weaning and then again at a set time (70 days is usual). Subtract the weaning weight from the 70-day weight and use this rate of growth measurement to select the fastest-growing lines, a very heritable trait. As an example, kit #1 weighs 1.2 pounds when weaned at day 30 and 5.4 pounds at day 70. It has a growth rate of 5.4 - 1.2 = 4.2 during this time period. Kit #2 was 1.8 pounds at day 30 and 5.7 pounds at day 70. Though it is heavier at day 70 than kit #1, its actual rate of growth during this crucial time period was inferior (5.7 - 1.8 = 3.9). Again, you want to make sure that you are comparing "apples to apples," especially in non-climate-controlled facilities. If you have several litters (which you usually will) all raised at the same time, however, it is fairly easy to pick out the lines among them that are producing the fastest kit rate of growth after weaning.

PRACTICAL SELECTION STRATEGY

The table on page 107 depicts a practical example of how to collect data for the most common selection criteria: fast kit rate of growth. In this case, 15 doe rabbits (of similar ages) (A-O) are bred with 5 buck rabbits (1–5) in such a way that each doe is rotated with each buck for a total of 5 breedings. The object is to determine which lines to keep as breeders based on their offspring's growth rate.

The entire litter from each mating is weighed at weaning (30 days) and then at harvest (70 days), with the weaning weight subtracted from the harvest weight to get just the weight gained from weaning to day 70. That number is then divided by the number of kits in the litter to get the average kit weight gain for each litter. This number is recorded for each doe/buck pair in the table.

Once the data is collected, the farmer decides on a cutoff number for does to be used for future breeding. In this case, the farmer decided 3.8 pounds or greater of kit weight gain from weaning to day 70 was desired. Each doe is then evaluated on how many of the five breedings her kits achieved a 3.8-pound rate of growth (or better). These numbers are recorded on the bottom of the table for the does and to the right for each buck. The highlighted numbers on the bottom row of the table show that eight does (A, C, D, F, G, H, I, and N) are superior in this criteria, as their kits hit this rate of growth at least three out of five breedings. Seven lines (B, E, J, K, L, M, and O) should be culled.

Next, the bucks are evaluated. Out of their 15 breedings, how many of their progeny reached the 3.8 cutoff point? This is recorded in the table's rightmost column but must be added up for each buck. The table on page 108 shows that bucks numbered 2, 3, and 5 had 9, 8, and 10 litters (respectively) reaching this cutoff point. These should be your future breeders. No kits from bucks 1 and 4 should be kept as brood stock.

Evaluation of the Commercially Important Trait of Kit Growth Rate

DOE LINES

		A	B	C	D	E	F	G	H	I	J	K	L	M	N	O	3.8
Breeding 1	1	**4.0**	3.2	**3.8**													2
	2				**4.1**	3.4	**3.8**										2
	3							**4.0**	**4.0**	**4.1**							3
	4										3.0	3.5	2.8				0
	5													**4.0**	**4.0**	2.8	2
Breeding 2	2	**4.1**	**3.8**	**3.8**													3
	3				**4.0**	3.5	**4.1**										2
	4							3.6	3.5	**3.8**							1
	5										**4.0**	**3.8**	3.0				2
	1													3.5	3.0	3.0	0
Breeding 3	3	3.5	3.3	**3.8**													1
	4				3.7	3.7	**3.8**										1
	5							**3.8**	**4.2**	**3.8**							3
	1										3.6	3.5	2.8				0
	2													**4.1**	**4.2**	3.5	2
Breeding 4	4	**3.9**	3.4	**4.0**													2
	5				**3.8**	3.2	3.5										1
	1							**3.8**	**4.1**	3.7							2
	2										3.5	3.5	3.0				0
	3													3.0	**3.8**	3.1	1
Breeding 5	5	**3.8**	3.6	**3.9**													2
	1				3.6	3.5	**4.2**										1
	2							**3.8**	**3.8**	3.5							2
	3										3.1	**3.8**	2.9				1
	4													3.6	2.8	3.0	0
	3.8	4	1	5	3	0	4	4	4	3	1	2	0	2	3	0	

BUCK LINES

Each doe (A-O) is bred to each buck (1–5) for a total of five breedings. The bucks are rotated so that each buck breeds three does on each breeding date and the does have a different buck every time. Their progeny's growth rate (measured from day 30 to day 70) is averaged for each litter, and this number is recorded in the table. Bold numerals are the litters that achieved a 3.8-pound growth rate or better. The far right column is the number of litters each buck produced that met this criteria (which is further analyzed in the following table). The bottom row is the number of litters each doe produced that reached this criteria. The highlighting on the bottom row shows the eight does that were superior, having three or more litters out of five breedings that reached this 3.8-pound goal.

Evaluation of Bucks for Future Breeders Based on the Number of Litters Reaching or Exceeding the Target Kit Rate of Growth

	Buck 1	Buck 2	Buck 3	Buck 4	Buck 5
Breeding 1	2	2	3	0	2
Breeding 2	0	3	2	1	2
Breeding 3	0	2	1	1	3
Breeding 4	2	0	1	2	1
Breeding 5	1	2	1	0	2
Total Litters ≥ 3.8	5	**9**	**8**	4	**10**

Five bucks (Bucks 1–5) are bred on five different dates (Breedings 1–5) with three different does on each date. The numbers of litters from each buck is recorded in the table in which the average litter growth rate between 30 and 70 days reached 3.8 pounds or greater. Bucks 2, 3, and 5 should be retained as breeding stock because over half of the 15 litters they sired exceeded the target growth rate. Bucks 1 and 4 should be culled.

Rabbit with temporary marker identification number in the ear, a good alternative to tagging or tattooing.

Once you have selected for the lines with the fastest growing kits, you can move to your next selection criteria.

MARKING AND IDENTIFICATION

To do any sort of selection of your stock, you will have to have some way to mark potential breeding stock so that they can be followed. In a small facility or family farm, tattooing the ears or ear tags are options. (We never used ear tags and saw a good deal of irritation of the rabbit's ear in stock that we bought from breeders who did use them.) Tattooing is a good, permanent identification, but in larger facilities that produce hundreds of meat rabbits per month, it is time consuming, as well as invasive and painful for animals that are not usually kept long-term.

As an alternative, we used various colors of long-lasting, all-weather, nontoxic livestock spray paint to temporarily mark kits we were evaluating

This rabbitry's setup includes tattoos in the adult's ears and clipboards on the top of each cage for recording the barn's records.

for possible future breeders. If we decided that a rabbit had met our selection criteria, we could identify it by the paint markings and move it to its own cage and begin a more permanent record. Once a kit was moved to its own pen as a breeder, we utilized a cage-card system where the identifying cage card with the rabbit's name or number on it is simply moved whenever the rabbit is transferred to a new cage—thus eliminating the need for any marking at all.

If you are only keeping your own replacement breeders or selling commercial, non-registered breeding stock, permanent marking is not essential. If you are selling several commercial breeders to an individual and you do not tattoo, it is common practice to use permanent markers to note a number in each rabbit's ear, then provide the buyer with a limited pedigree and degree of relation between animals to help them avoid inbreeding among the animals they have purchased. These markings will only last a few days, however, and the buyer must then use whatever identification system they choose.

No matter what system of marking you use, the identification of each rabbit should help you to classify that animal in ways that will help you with your operation. Often, the particular maternal or paternal line will be indicated by a letter. For example, if the offspring is the 36th animal kept from maternal line B and paternal line F, the offspring may be identified as BF36. If you wish to keep track of birth years as well in the identifier, you might have 2016BF36.

If you are providing actual pedigrees and/or registration when you sell breeding stock, you will need to permanently mark an individual with an ear tattoo (and charge for the added time and expense). It is customary to put farm tattoos in the left ear and save the right for the formal rabbit breeding association's registration numbers.

Often, a farm identification is included in such a tattoo to help prevent other breeders from passing stock off as yours (such as using CR48BA for Chigger Ridge rabbit #48 out of rabbit lines B and A). This animal will then be put in your breeder records, along with to whom and when you sold it.

Whatever naming or numbering system you choose to use, records of each animal should be kept on their parentage to prevent inbreeding and on their performance to select future breeders.

RECORDS

Just what information should be kept in a rabbit's records to help with selection of desired traits and avoidance of problems? The most obvious is the ancestry, through at least 5–6 generations if possible, but there is much more in the way of data that will be necessary.

First, you should record any time that a doe or buck has a problem. That is anything, whether it is ear mites, lack of appetite, respiratory problems, diarrhea, sore hock, overgrown teeth, refusal to breed, weight loss, kindling difficulties, or litter loss. You will need to meticulously record which animals are bred to which and what the results were. Your records should answer the following questions:

- How many kits were born from the breeding?
- Did the doe make an adequate nest?
- Were there any diseases, abnormalities, or deaths in the litter?
- What was the birth weight of the litter?
- What was the weaning weight of the litter?
- What were the finishing weights of the kits?
- How long did it take to reach finishing weight?
- How soon after kits were weaned did the doe regain weight and rebreed?

All of this data must be analyzed in the context of weather conditions if you do not have a climate-controlled facility, so this should also be recorded.

If you consistently maintain this data, you can set up comparison tables such as the preceding example (see Practical Selection Strategy) to isolate and select for whatever trait you are looking to improve. It does not take as much time as it sounds to record and interpret data, and it is part of being a successful breeder. Rabbit raising, with the fast time between generations, takes more effort in this area than many other livestock species. Having such records will also greatly increase your future breeding stock sales and allows you to charge more for your "selected breeders" than other rabbit raisers (since you will have the statistics to back up your claims).

WHEN TO CULL

Rabbits should be culled when they:

- Become nonproductive
- Don't meet your selection criteria to improve your stock
- Have health issues that limit their quality of life or threaten other animals in the facility

Remember that rabbits in the wild rarely live past a single year. They are designed by nature to grow fast, reproduce when young, and then die to provide food to the multitude of predators that rely on them as a food source. Thus, older rabbits (4–5+ years), by their very nature, become less productive. The bucks will be less inclined to pursue a reluctant doe for breeding, and a doe will begin to have fewer kits, and her lactation will decrease. When this point is reached, it is time to have the cage space filled by animals that will be able to produce. Likewise, if you have a very promising young doe ready for breeding from lines that are showing the traits that you need to improve in your stock, it is sometimes necessary to cull an inferior breeder to make room for the new one.

As in the wild, these older or less productive animals can still provide nutritious food for yourself or your friends and family. Though you can't sell older animals as "fryer" rabbits, they can still be used as "stewers," providing a wonderful meal in the Crock-Pot. There are certain buyers that actually prefer to buy the larger animals if they are priced less per pound than a fryer. Sometimes there is also a market for larger rabbits in the pet food industry, and the skins of older animals are more valuable in rabbit fur industries (fryer-age skins are considered too thin for all but craft uses).

> **Though you can't sell older animals as "fryer" rabbits, they can still be used as "stewers."**

CHAPTER 7

SLAUGHTER AND CARCASS

Disclaimer: The slaughter methods mentioned in this chapter may cause risk of injury or death to the operator if done without proper training and safety considerations. These are just suggestions of possible methods of slaughter; it is up to the reader to obtain training, maintain proper equipment, check on state or federal regulations, and follow safety protocols.

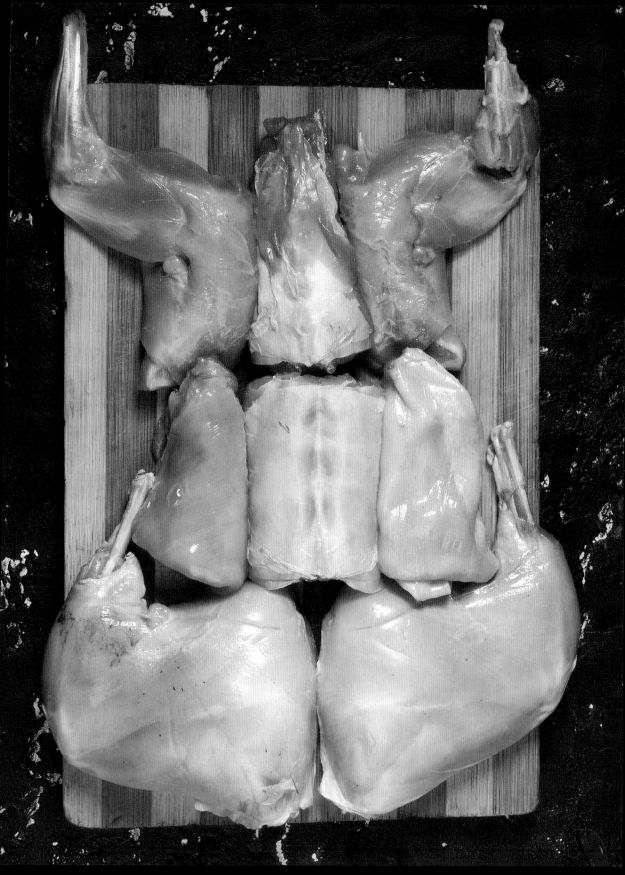

SLAUGHTER METHODS

Unless you live in one of the few states that allows on-farm slaughter, packaging, and sale of rabbit to the public, chances are that most of the slaughter from your farm operation will be done by professionals at a meat processor. However, the small family operation raising animals for its own consumption may have to slaughter its own rabbits, as a meat processor will likely charge too much to be economical.

Though there are no actual laws governing the most humane way to carry out rabbit slaughter for personal consumption, or for euthanasia of ill animals, there *are* laws (which vary from state to state) regarding humane slaughter of rabbits *for sale to the public* as meat (see Chapter 8). Humane slaughter entails the least possible stress to you and the animal. Slaughter of animals that you have raised is never a pleasant task but is part of any meat animal operation.

> To bleed out a rabbit safely and humanely, the animal must be rendered unconscious first.

In the slaughter of a rabbit for meat consumption, death occurs when the animal is cut across the neck, severing the carotid arteries. It is then hung up and allowed to "bleed out," which gives the most palatable meat product. A rabbit that is not bled will taste more gamey and will not preserve as well. To bleed out a rabbit safely and humanely, the animal must be rendered unconscious first. Below, we will briefly describe a few options you might consider to accomplish this. The reader must perform further research on any slaughter technique and the methods that are proscribed by your particular state for meat sales to be sure he or she is in compliance with any state regulations.

Using a firearm (usually a .22) to dispatch a rabbit is quick but does not allow for bleeding out. It is usually used for animals that are ill and will not be eaten. You must obviously be familiar with the safe use of a firearm to utilize this method.

A second method of euthanasia is injection of euthanasia drugs. This is considered a painless death but can only be administered by a veterinarian. It is expensive, and the animal cannot be eaten afterward.

Cervical dislocation (breaking the neck) is a common way to euthanize a rabbit before bleeding it. This is accomplished by immobilizing the head and pulling the back legs sharply and strongly. Rabbits greater than two pounds, however, may have neck muscles too strong for many people to get a clean, quick kill. There is equipment available, such as the Hopper Popper (www.theoriginalhopperpopper.com), that creates the proper angle and leverage so that larger animals can be dispatched more easily. Immediately following cervical dislocation, the animal should be bled.

There are also a number of "stun guns" that can be purchased to allow for stunning the animal prior to bleeding out. (These are not the same as TASER® stun guns used by law enforcement—there is no electrical shock.) The objective of this tool is to produce a concussion in the brain that renders the animal unconscious before bleeding them out. These guns replace the older, less reliable method of using a hammer to stun the animal. When done with appropriate force and accuracy, the hammer can be humane, but if a person does not perform the blow correctly, hits the wrong area, or does not use enough force, the animal may feel pain. The stun gun decreases the possibility of an inhumane death. As soon as the animal is stunned, it should be bled out.

Rabbit liver is an indicator of animal health and will be checked by the meat inspector. It should be smooth and without blemish. (It is also larger and sweeter than chicken livers and prized by chefs.)

Gassing is another method of rendering an animal unconscious before bleeding out. Carbon dioxide (CO_2) is currently used in the swine and chicken industry in some parts of the world and is considered in those locations to be a humane form of anesthesia before slaughter. There are four phases to CO_2 gassing:

> **Humane slaughter entails the least possible stress to you and the animal.**

1. Almost immediate loss of consciousness
2. An excitation phase with rapid breathing and uncontrolled movements, but where the animal is unconscious
3. Deep anesthesia with total relaxation and loss of eye reflex and pain reflex, but where breathing and circulation remain intact
4. Death from lack of oxygen to the brain

If an animal is removed from CO_2 exposure even during phase 3, it will recover completely in under a minute. Therefore, if bleeding is desired, it must be slaughtered *immediately* after removal from CO_2 exposure at stage 3. The second phase, when the animal is moving, can be difficult to watch, even for those knowing that the animal is not conscious. CO_2 chambers should have a clear side or top so that animals can be monitored and removed during phase 3 when the animal is relaxed, unconscious, has no pain reflexes, and can be easily bled out.

A compressed-gas CO_2 cylinder can be rented on a monthly basis from gas companies and swapped out when they are empty. You will need a regulator and hoses to control the flow of carbon dioxide into the gassing chamber. Dry ice is a form of solid CO_2, and when small pieces are placed in water, they will rapidly turn back into a gas. Some farmers use this instead of compressed-gas cylinders to create CO_2 gas for rendering a rabbit unconscious before slaughter (although it is more difficult to regulate the amount in the chamber and ensure rapid passage through the phases to stage 3). Care must be taken that the animal does not come into direct contact with the dry ice, as it is extremely cold and can cause pain.

Both dry ice and compressed-gas CO_2 cylinders should be handled only by those who have had training in safe use and storage. Dry ice can cause frostbite. It should be handled with gloves and goggles. If compressed-gas cylinders are damaged (say from tanks being knocked over and the safety valve broken), explosions can result. (They should always be chained to immobile objects to prevent this.) CO_2 should be used with extreme caution if used indoors. CO_2 is heavier than air. If there is a leak in the chamber, children and pets located near the floor could be at risk from CO_2 buildup in an enclosed, poorly ventilated room.

CO_2 can cause immediate unconsciousness in people who are overexposed. This author has firsthand knowledge of this after having passed out by sticking my head—just for an instant—too far down in a dry ice storage bin. This is why I believe the animal is indeed completely unconscious and has no memory during the later phases of CO_2 exposure (even though there is some literature that states CO_2 may be less humane than previously believed). If a person passes out from carbon dioxide exposure and is not removed to fresh air, this can result in death.

> **Reducing the animal's anxiety before and during slaughter is not only part of ethical farming, it will also produce a higher quality meat product.**

We do not ever recommend using CO (carbon *mono*xide) gas as a method of euthanasia! We have heard of farmers who use car exhaust fumes piped into a sealed chamber to try to render a rabbit unconscious before slaughter. This gas is not the same as CO_2! People overexposed to CO_2 will recover without injury if removed in time to fresh air, but carbon monoxide will build up in the blood and replace the oxygen attached to red blood cells. Humans exposed to carbon monoxide can end up with organ damage, including brain damage, and can easily die even if removed to fresh air. It is *far* more dangerous than CO_2.

Whatever method you choose for slaughter, it is just as important (besides it being quick and as painless as possible) that the stress levels and anxiety of the animal be taken into consideration before the procedure is performed. All animals will become fearful at the smell of blood, and thus it is recommended that whatever chamber, table, or apparatus that is used be cleaned between each animal that is dispatched.

It is also recommended that animals waiting to be slaughtered be kept in a separate area to prevent any distressing sounds or smells from reaching them. Unfamiliar animals should not be mixed in one cage before slaughter but kept separate to prevent tension and fighting. Of course, it goes without saying that the animals should be in a comfortable environment with cold or heat stress avoided. Reducing the animal's anxiety before and during slaughter is not only part of ethical farming, it will also produce a higher quality meat product.

TRANSPORTING TO A PROCESSOR

The same consideration should also be given to animals that you are transporting to a slaughter facility. Using group cages is fine, but the sexes should not be mixed together. This may result in attempts to

Protection from the elements and from stress are important during transport. Aside from being inhumane, stressful conditions will lead to an inferior product.

mate, producing stress and fighting. If cages are to be stacked for transport, the lower cages need to be protected from urine or manure from the upper animals. The outer cages should be protected from wind in extreme cold. When transporting in high temperatures, you should transport fewer animals in the middle and lower cages to prevent heat stress. All animals should be protected from precipitation in cold weather; however, light misting before transport helps rabbits transported in hot weather to stay cool.

Many slaughter facilities will request food or water withholding for a certain period of time before slaughter, for either meat quality or ease of processing. There has been some research on the effects of this withholding on carcass quality for rabbits. The results indicate that withholding of feed for 16–18 hours but no withholding of water produced the best results (Xiong et al. 2008). You need to work with your processor and find out exactly when they plan to process your animals to determine when to remove your feed from your rabbits.

We have known a few rabbit farmers that try to save money by withdrawing feed 24 hours or even longer before transport to a slaughter facility. Remember that one day of not eating for a rabbit can translate to a human not eating for 3–4 days. This amount of feed withdrawal may cause systemic stress, possibly resulting in death during transport. Even if this does not occur, the carcass quality will be affected as fat stores are utilized.

These recommendations for humane preslaughter treatment and transport are not just for ethical considerations but are important for meat quality. Poor meat quality (mushy, dry, or tough) in rabbits is known to occur from preslaughter adrenaline rush or muscle-meat pH changes resulting from various stressors.

CARCASS QUALITY

According to the USDA:

If it is over **10 weeks** or **5.5 pounds**, label your product as simply "domestic rabbit" and not "domestic *fryer* rabbit."

1. A fryer rabbit is less than 10 weeks old and is 3.5–5.5 pounds live weight
2. A roaster is 10 weeks to 6 months old and is 5.5–9 pounds live weight
3. A stewer is over 6 months and is a minimum of 8 pounds live weight

That being said, the industry in practice tends to disagree. Every meat processor we worked with, and every live rabbit buyer we sold to, considered a fryer to be any rabbit under four months and 6.25 pounds maximum live weight. (Over this cutoff in either age or weight was considered a roaster/stewer). We found rabbits even at the high end of this industry standard have tender, fine textured, pearly white meat, and a great meat-to-bone ratio. (We therefore respectfully disagree with the USDA definitions.) For legality's sake, if it is over 10 weeks or 5.5 pounds, label your product as simply "domestic rabbit" and not "domestic *fryer* rabbit."

The rabbit "backstrap" (saddle or loin) is the most tender and highest valued cut on the rabbit. The term "long in the saddle" refers to any meat animal with more flesh in this important area. Most of the meat of the rabbit is on the hind legs—you want a carcass with solid, heavy thighs. However, the front legs should not be neglected—they produce very yummy cuts as well (similar to chicken wings)—and the shoulders should be deep-fleshed. A "Grade A" fryer rabbit will also have a fair amount of fat in the interior body cavity.

The photo panel on page 119 illustrates of the importance of carcass quality and slaughter weight of a rabbit. To the right of each pairing is the desired carcass of a high-quality product—a domestic rabbit raised for sale to the public at restaurants, farmer's markets, or high-end grocery stores. To the left in each pairing is a cheap, imported (in this case from China) rabbit sold at many chain-store groceries. These imported rabbits are typically slaughtered at a much lighter weight than we recommend and are not as highly selected for carcass superiority. You can see in the photos that the imported rabbits are not nearly as wide in the loin and have less flesh along the back, hips, and thighs. They have very poor front leg and shoulder muscling and almost no fat.

This 1.4-pound Chinese imported rabbit was purchased at a grocery store for $10 ($7.14/pound). The US domestic rabbit, at 2.9 pounds, was purchased for $20.30 ($7.00/pound). When cut up and the waste areas of the neck, chest, backbone, and pelvic bone discarded, the imported rabbit meat weighed only 0.4 pounds (so actual cost of meat is $25 per pound). Whereas the US domestic rabbit meat weighed out at 2.1 pounds (so actual cost of meat is $9.66 per pound). The final cuts weighed were: bone-in front and back legs, deboned loin, tenderloin, and

To the right of each pair is a Grade A American domestic meat rabbit with wide loin, excellent muscling, and with lighter colored fat visible over the shoulders and in the body cavity. To the left is a poorer quality imported rabbit which has a narrow body, is not well fleshed, and has no fat at all.

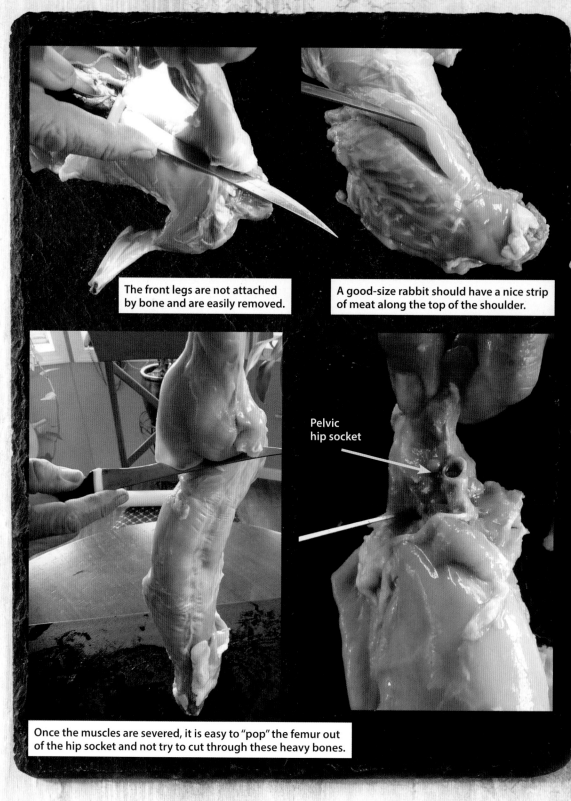

The front legs are not attached by bone and are easily removed.

A good-size rabbit should have a nice strip of meat along the top of the shoulder.

Pelvic hip socket

Once the muscles are severed, it is easy to "pop" the femur out of the hip socket and not try to cut through these heavy bones.

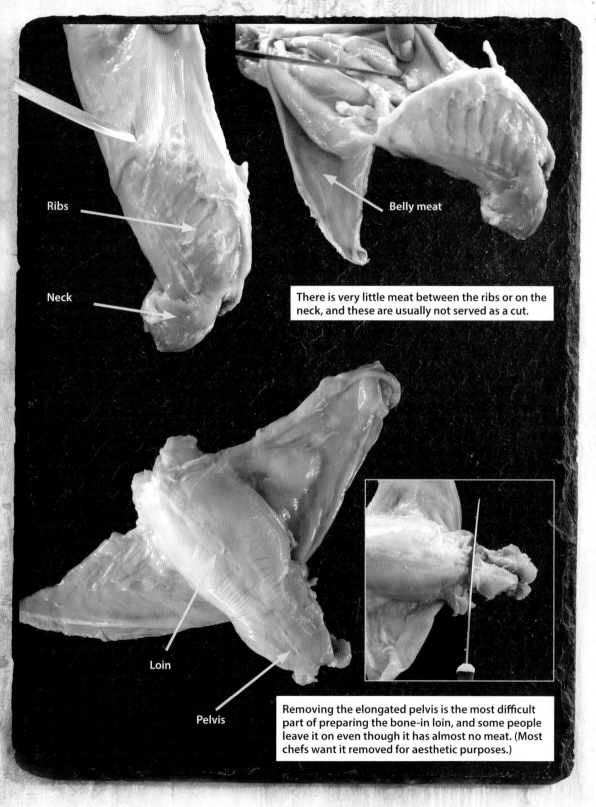

Ribs

Neck

Belly meat

There is very little meat between the ribs or on the neck, and these are usually not served as a cut.

Loin

Pelvis

Removing the elongated pelvis is the most difficult part of preparing the bone-in loin, and some people leave it on even though it has almost no meat. (Most chefs want it removed for aesthetic purposes.)

shoulder strips. It is obvious that the more muscular US domestic rabbit is more economical as far as price per pound of actual meat.

RABBIT CUTS

Although rabbit is excellent cooked whole in an oven or stew pot, other cooking methods such as frying or barbecuing work better by dividing the carcass into cuts. This is actually very easy to do and requires only a single sharp knife. The first step is to slice off the front legs, which are not attached to the body by bone and are very easy to remove. In a good-size rabbit, there will also be a nice strip of meat along the top of the shoulder. This can be taken as part of the front leg, or it can be cut off separately (page 120, top).

Once the front legs are removed, the back legs are next. Many people make this much more difficult than it needs to be. The hind leg has a thick femur (thigh) bone with a rounded top (hip joint) that is seated deeply in an elongated pelvic bone. You will encounter these heavy bones if you try to cut off the legs straight across with a knife or cleaver. Instead, hold the rabbit up by one hind leg and allow the body to hang down. Then, run the knife round and round the top of the leg where it meets the pelvis. Once all the muscle and ligaments are severed, you will be able to see the rounded head of the femur begin to pop out of the pelvic bone. It is an easy matter at this stage to detach the leg with a knife or just twist the hind leg until it comes all of the way out of the pelvic socket (page 120, bottom).

After all four legs are detached, it is time to go after the highest-end cuts: the loin and tenderloin. The loin is on the *top* along both sides of the backbone (called the backstrap in a deer). It is much larger than the tenderloins, which are *inside* the body cavity along the backbone. In a rabbit that is too light at slaughter, the tenderloins are almost too small to cut out, but in a larger rabbit, they are a couple of nice bite-size pieces that are very tender. The loins can be left attached to either side of the backbone or can be cut off and served boneless. For a "bone-in" loin, you need to remove and discard the rib/neck region and the pelvis (these areas have very little meat).

To prepare this cut, look for the bottom rib as shown in the top of the photo panel on page 121, then trim along the tips of the ribs and lay out the thin belly meat (the rabbit bacon), leaving it attached to the loin and backbone. This belly meat should look like a bird wing when stretched out. Do the same on the other side so you have two "bird wings." At this point, the chest area with ribs and the neck is only attached to the rest of the rabbit by the thin backbone and can be removed and discarded by simply cutting through the backbone between the vertebrae or just rocking it to and fro with your hands until it snaps.

Removing the pelvis from the loin is the most difficult part of the butchering. Cut around the top of the pelvis until you find the thin backbone, and then cut between the vertebrae or snap it by rocking it between your hands back and forth until the backbone breaks. The pelvis in a rabbit is very long, and the object is to get up far enough to clear the top of the pelvic hip bones without losing too much of the loin meat. When the pelvis is removed, you are left with a single whole bone-in rabbit loin, called "rabbit saddle," with bird wing belly meat still attached to each side.

Butchering for deboned loins (as opposed to the single bone-in loin cut described above) provides two cuts of loin instead of one, allowing more people to be fed from this valued area. This process simply involves cutting the loin from the backbone at both sides of the spine. Make a slice straight

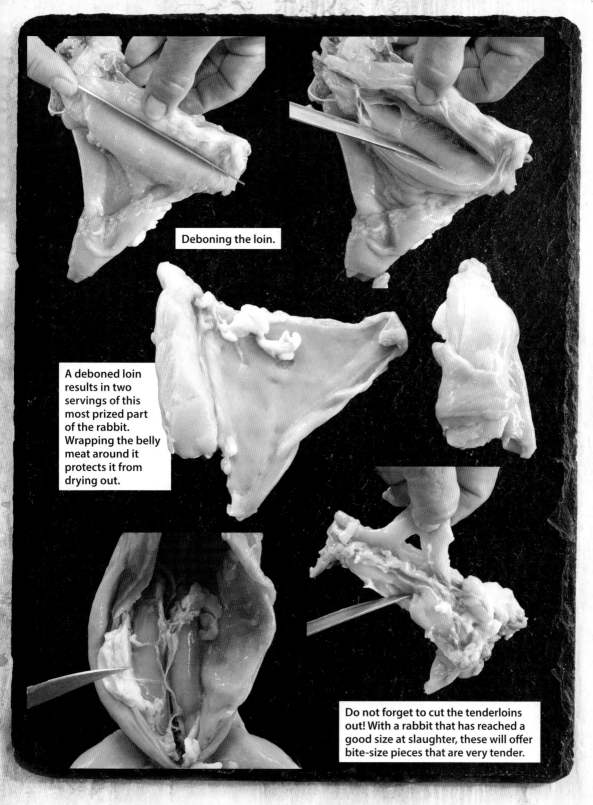

Deboning the loin.

A deboned loin results in two servings of this most prized part of the rabbit. Wrapping the belly meat around it protects it from drying out.

Do not forget to cut the tenderloins out! With a rabbit that has reached a good size at slaughter, these will offer bite-size pieces that are very tender.

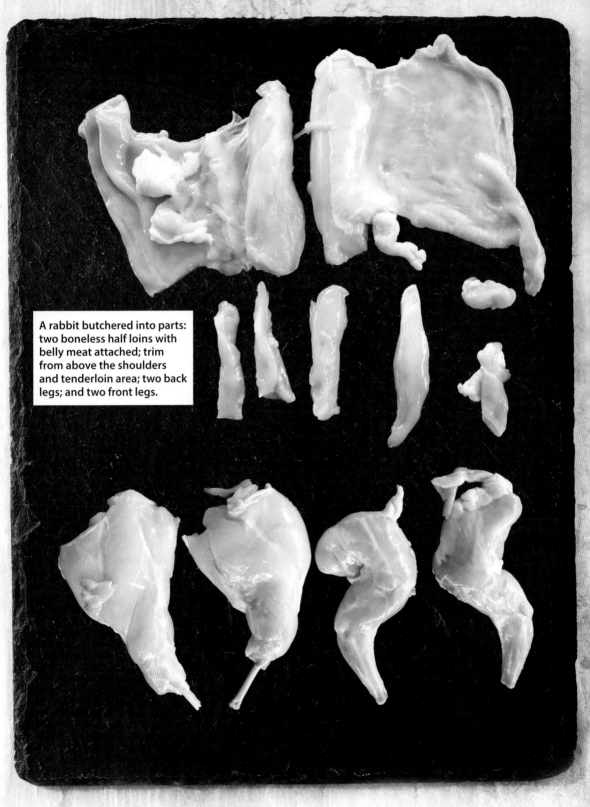

A rabbit butchered into parts: two boneless half loins with belly meat attached; trim from above the shoulders and tenderloin area; two back legs; and two front legs.

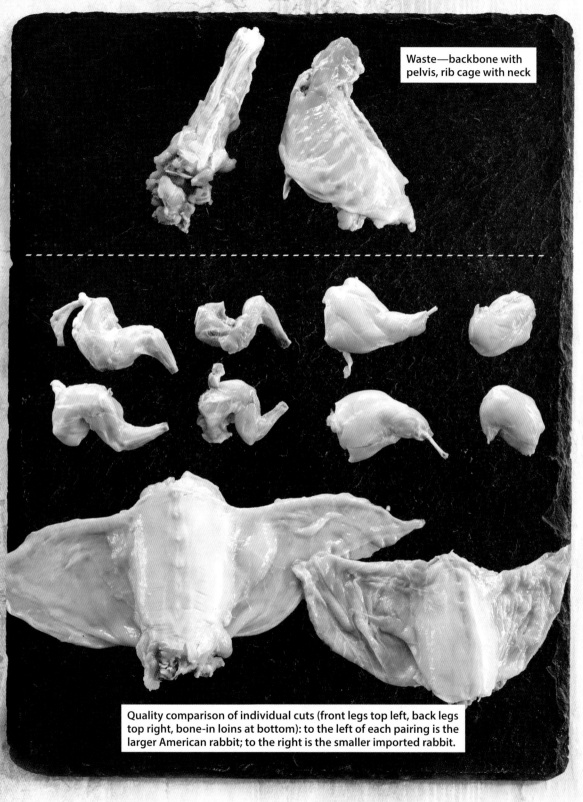

Waste—backbone with pelvis, rib cage with neck

Quality comparison of individual cuts (front legs top left, back legs top right, bone-in loins at bottom): to the left of each pairing is the larger American rabbit; to the right is the smaller imported rabbit.

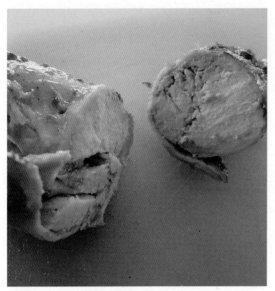

Barbecue boneless loin "roll ups" with the extra meat trim cooked inside!

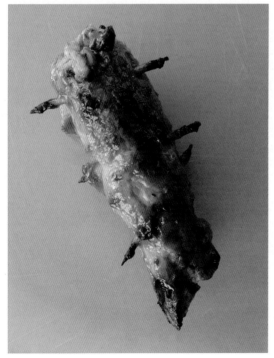

Wrapping the belly meat around the boneless loin protects this tender cut from drying out during grilling.

down the back, and slide your knife down along the side of the spine to remove the loin with the rabbit belly meat "bird wing" still attached. You can cut off and cook the belly meat separately or (more usually done) wrap it around the loin as shown on page 123.

Finally, do not forget those tenderloin strips from inside the carcass. In the bottom left of the photo panel on page 123, you can see these small, long strips at the tip of the knife in the body cavity before the start of butchering. The photo to the bottom right of this same panel depicts removing them from the backbone during the final stage of butchering, when they are easily reachable without tearing them.

You now have your rabbit completely butchered, resulting in two front legs, two back legs, two boneless loin cuts with belly meat, and several small strips of meat from above the shoulders and the internal tenderloins (see photo page 124). The waste is the backbone, pelvis, chest ribs, and neck (see top of page 125). When cooking a rabbit whole, these waste areas will provide a bit of additional meat but are not cuts you usually serve fried or grilled to anyone, as they are mostly bone.

On the bottom portion of the photo panel on page 125 is a comparison of high-quality American domestic rabbit (on the left in each pairing) compared to low-weight, imported Chinese rabbit cuts (on the right in each pairing). Front legs are on the upper left, back legs on the upper right, and whole loins are on the bottom.

COOKING RABBIT

There are a million different rabbit recipes available in cookbooks and on the Internet, and we cannot begin to do this subject justice in this book. Since rabbit has been raised or hunted and eaten in nearly every culture around the world for centuries, there are countless options for rabbit meals with a German, Italian, French, South American, Greek, Russian, Chinese, U.K., and of course, American flavor, to name but a few. Whether you like spicy Cajun, Creole, or Mexican flavors, a white wine sauce with garlic and rosemary, a red wine sauce and mushrooms, conventional barbecue flavoring, or—one of our favorites—sesame ginger, all of these and more will complement rabbit.

You will notice, perhaps, that many of the recipes you find for rabbit are stews. This is because wild rabbits may be of any age and tenderness and will have less fat and a more gamey taste than a domestic farm-raised rabbit. Recipes that have been handed down for wild rabbit call for brining the carcass before cooking, slow stewing, heavy marinating, or adding fat such as bacon.

These traditional recipes are great but are not a necessity for a good domestic rabbit. A quality, young, farm-raised rabbit can be fried, grilled, even deboned and sautéed, as well as stewed or slow roasted with or without marinades—in fact, it can be cooked any way you cook a fryer chicken. Since there is no skin, you have to be careful not to overcook it if not cooking in liquid, but rabbit has a dense meat that tends to stay juicy even without the skin.

A good 2.5–3-pound rabbit carcass can easily feed four people when cut as described earlier. If it is cooked whole, there will be more available meat, and it will usually feed five people. If cooking for two, a good option we found is to stick a rabbit in the slow cooker or roast it whole for the evening meal, and then pull all the remaining meat off the bones and serve it with a gravy or cream sauce the next day over toast, noodles, or rice for breakfast or lunch.

Leftover rabbit can be pulled off the bone, mixed with gravy, and served the next day over rice, noodles, or bread for a change of pace.

"GOING COMMERCIAL" AND REGULATIONS

COMMERCIAL CONSIDERATIONS

This next section of the book deals with various aspects of the commercial rabbitry. By this, we mean everything from the small rabbit farmer raising just enough surplus animals to sell as breeders or as meat to cover their own food-raising costs, to the 1,000-kit-a-month full-scale rabbit agricultural operation.

While the previous sections of this book are applicable to both the hobby and commercial rabbit raiser, these final chapters will deal more with how to run a *profitable* agricultural enterprise. We will discuss venues for the marketing of meat rabbits, the laws and regulations that you may encounter in the United States while engaging in rabbit commerce, and how to determine your prices. For our international readers, while we cannot include all the government regulations you might face, the *process* for going about moving to a commercial operation is the same everywhere.

There are three questions that you must answer for your own particular area before you consider any commercial operation:

1. Calculating the distance and availability of possible markets
2. Ascertaining the laws governing live rabbit sales, meat sales, and meat processing in your state.
3. Determining the availability of rabbit meat processors or packers who meet your state's regulations if on-farm processing is not allowed in your locale. (If on-farm processing *is* allowed, you must work out how it is regulated and inspected.)

> **The biggest bottleneck in the growth of the US meat rabbit industry is a lack of federally inspected processing facilities.**

We will deal with how to go about wading through these three initial questions in this chapter to determine if "going commercial" is a viable option for you. Then, in the following chapters, we will introduce specific marketing techniques for the various outlets for your kits and the cost analysis needed to set your rabbit prices to realize a profit.

MARKET OUTLETS: LIVE RABBITS

One of the easiest ways to sell your rabbits is through a contract with a live meat rabbit buyer. Your State Department of Agriculture may be able to tell you if there are any meat packers that operate in your area who might buy your rabbits live. Some of these meat processors may even be willing to send a "rabbit runner" to a set point where several farmers bring their animals—so don't count out rabbit processors that are more distant until you contact them. The American Rabbit Breeders Association (ARBA) and other rabbit groups may also have a list of processors—some of which may buy live meat rabbits directly. (While you are researching this, find out if the processor requires any particular husbandry protocols while you raise them.)

These are not the only live rabbit buyers out there, however. Sometimes other farmers with established markets cannot raise enough rabbits to meet their market demands and may be willing to buy your live rabbits to process along with their own. Check if other farms in your state are selling rabbit, and contact them to see if they need an additional rabbit grower to provide for their customers.

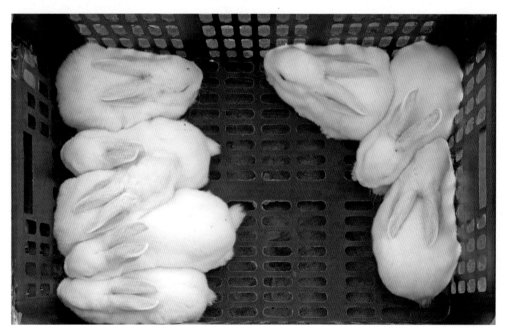

If you are not lucky enough to have a meat processor who will buy your rabbits on contract, you can sell them at many livestock auction barns.

Another option is selling at livestock sale barns. Many of these livestock markets will sell live rabbits for you at auction—just like they do with cattle, sheep, or goats (usually on days that they do other small animals, such as chickens). You can also sell live rabbits from a booth at some flea market or farmer's market venues. It should be noted, however, that fewer and fewer Americans are willing to slaughter animals for food. Luckily, many immigrant populations are very happy to find an available source of meat animals and disdain frozen prepackaged meat. You need to look at the demographics of these local markets before you count on them as an outlet.

Breeding stock sales can be a lucrative part of your business.

Selling quality breeding stock to other rabbit raisers is also an outlet for live rabbits. Not every animal should be used for breeding, and you will still need an outlet for the rest of your crop. Another potential source of live rabbit buyers are the 4-H and Future Farmers of America (FFA) programs. Because of the lower cost of purchase and housing and the fact that there are few restrictions for rabbits in most communities, rabbits can be a very rewarding tool for teaching youth about sound and sustainable farm practices. If you like kids and are willing to volunteer some of your time teaching, you may develop a profitable and very rewarding relationship with your local school chapters. This may be but a portion of your rabbit market outlet, but it can be used as an educational tool to increase the overall demand for rabbit meat in your area.

MARKET OUTLETS: RABBIT MEAT

Though specific marketing ideas will be presented in the following chapter, you must evaluate the *potential* for rabbit meat sales before you buy your first rabbit—especially if you are thinking about a larger-scale operation.

Rabbit meat can be marketed directly to the public through farmer's markets. This will require you to be available for most weekends throughout the normal growing season at least. Some markets in larger cities are even open every day. You should visit your local markets and inquire about booth space costs and availability, electricity for freezers, and market regulations. Then you need to observe the traffic flow. Find out if there is an immigrant population that might be looking for rabbit. Discover if there is a trend toward antibiotic-free, naturally raised, "exotic" meats in the upscale buyers. Determine if there are other sellers that might be competition or already attracting people to the market. Talk to these vendors if possible and find out what kind of sales volume they might have in a day.

Rabbit meat can also be marketed through restaurants. The "chain store" type establishments with the same menu throughout the country are not usually an option. Look for a farm-to-table operation or an upscale, high-end business that offers exotic cuisine. Buffet-style venues may be interested in doing a "Rabbit Tuesday" or some such marketing ploy to bring in more business on slower days. We will discuss more in Chapter 9 on how to approach these businesses.

Be aware that many restaurants have their food delivered, and this may be a service you will need to provide, so look for establishments within reasonable driving distance. If you ship your product to restaurants, this may increase your costs to the point where you must set your prices too high for the market.

Rabbit can also be marketed through food stores. Again, it is more advantageous to look for local, individually owned places rather than the large chain grocery stores—especially as you just begin to build your rabbit herd. Small specialty organic or health food markets may be very interested, and their clientele willing to pay a fair price. However, even the larger chain stores (unlike the chain restaurants that must keep to specific menus) may be interested in stocking rabbit meat for something different to offer their customers.

Internet-based sales of rabbit meat throughout the United States is a possibility if you are a larger operation. To sell across state lines, you will definitely have to have each animal inspected at slaughter and a USDA or Food and Drug Administration (FDA) stamp of approval for each carcass (these stamps are described in the section below). You will also have to determine shipping costs closely, as you will likely encounter hazardous materials charges for shipping frozen product in dry ice.

Pet food sale is another outlet for rabbit meat. There is a strong movement in the United States toward feeding our dogs and cats more natural foods, especially allergy-prone animals. Some veterinary clinics or pet stores may carry your rabbit or be willing to promote you for on-farm sales to their clients that need natural food. Larger rabbits, such as culled breeders or smaller runt rabbits from a litter, may bring a good price as pet food sold directly to the public (with appropriate labeling) or to pet food companies. By-product sales, such as rabbit liver, heart, and kidneys, are popular pet food items. The sale of excess frozen "pinkie" kits for snake food is a possibility (usually marketed through pet stores). In our opinion, no rabbit, unless it is sick or diseased, should be wasted. Not only does this help your bottom line, it is part of the creed of sustainable agriculture.

Rabbit meat is becoming popular in pet food too.

PET RABBITS VERSUS MEAT RABBITS—WHAT'S THE (LEGAL) DIFFERENCE?

Rabbits intended for food are not regulated under the Animal Welfare Act (AWA), whereas rabbits intended for pets or research are. The AWA mandates specific housing, husbandry, and transportation requirements. There are many "fine points" you must obey, and your farm will be randomly inspected by the Animal and Plant Health Inspection Services (APHIS) without notice. You must not only be in compliance on that day, but have documentation and set farm protocols to prove that you stay in compliance year round. The husbandry requirements for selling pet rabbits are found in what is termed "The USDA Blue Book," and you can access the requirements following the link on page 188 under Animal Welfare Act and Animal Welfare Regulations from USDA, APHIS.

Any rabbit of any breed can be used for meat, and any rabbit of any breed can be a pet. Exactly how, then, do the federal authorities decide if you are selling pet rabbits and should therefore be licensed, inspected, and following Animal Welfare Act standards, or if you are selling meat rabbits and agricultural breeding stock and are considered a farmer and not subject to the AWA? Fines for violation of the AWA can be very high, so you cannot ignore this subtle difference!

So how do you know if you need to be licensed under the AWA? Basically, it involves your advertising, where you sell your rabbits, and your *intent* that determines if you are raising pets or meat rabbits. If you are wholesaling rabbits of any breed or size through pet stores, you definitely need to be AWA licensed. If you wholesale them to be resold through any retail stores (for example, at your Farmer's Cooperative for Easter), you will need to be licensed. If you sell live rabbits sight-unseen through the Internet, you must be AWA licensed.

However, if you sell them face to face at your farm as pets and you, the end-buyer, and the rabbit are all present, you may (or may not) be able to get an exemption as a "direct sales" pet breeder. The reason I say "may not" is that if you sell pets through *any* other venues as well, you may lose your direct sale exemption. The APHIS requires that you either hold a license or not; you cannot have only *some* sales be AWA exempt.

A license is required if the person:
- Has more than four breeding does and sells rabbits wholesale
- Has more than four breeding does and sells rabbits retail sight unseen
- Sells rabbits or rabbit parts to research institutions
- Has rabbits involved in promotional exhibits

A license is not required if the person:
- Sells rabbits at a place where the rabbit raiser, buyer, and animal are physically present at the same time
- Has gross sales of rabbits under $500 per year
- Sells rabbits for food, fur, or fiber
- Sells rabbits for agricultural purposes

The USDA "Blue Book" provides details of AWA animal husbandry requirements.

Meat rabbit farms do not need to be licensed under the Animal Welfare Act, but if you sell pet rabbits you must be registered under this law and comply with its regulations.

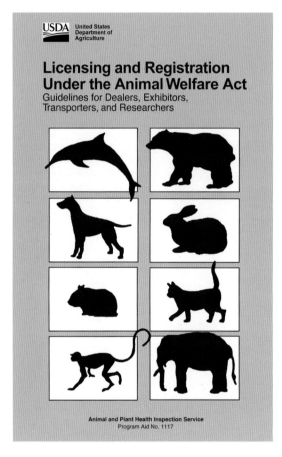

USDA United States Department of Agriculture

Licensing and Registration Under the Animal Welfare Act
Guidelines for Dealers, Exhibitors, Transporters, and Researchers

Animal and Plant Health Inspection Service
Program Aid No. 1117

The USDA "Gray Book" provides information on licensing and registration under the AWA.

- Maintains eight or fewer rabbits and exhibits them to the public
- Takes part in showing rabbits at agricultural shows or fairs

Basically, if you use the word "pet" on your website or social media, on a sign outside your farm, in a classified advertisement, on a bulletin board, or in conversation with prospective buyers, you may need an AWA license. Check with the USDA—before they check on you! (Only if you are a very small farm with less than four doe rabbits and sell less than $500 worth per year would you be considered a hobby breeder and exempt as such).

If you are still unsure whether you need a license, the APHIS has recently created a really nice self-service licensing assistant that can be used *anonymously*. On their site, you can answer questions regarding your breeding and sales activity and it will tell you whether or not you might need a license as a breeder, dealer, or exhibitor. If you need to be licensed, the application fee is $10, and then a yearly fee (currently between $30 and $750) is charged depending on your dollar volume of sales.

If a federal authority who has responsibility for enforcing Animal Welfare Act regulations hears or sees, by any means, that you are possibly wholesaling pet rabbits (rather than meat rabbits) without holding a license, he or she is required to investigate—and will! If you are found to be engaging in activity that requires a license, you will be given a chance to obtain one, and you will have 90 days after applying to correct any deficiencies on your farm to meet AWA standards.

So how do you stay "safe"? First, if you actively sell pet rabbits, obtain a license. Do not try to get around the law—the costs are just too high for violations. If you do not sell pets and are a

Wholesaling pet rabbits means you must be a licensed AWA facility, and you will be periodically inspected—without prior notice—to make sure you are in compliance with all husbandry protocols.

legitimate meat-rabbit farm, make it clear when talking with *any* rabbit buyer that you raise meat rabbits (not pets) or that you sell breeding stock to agricultural enterprises. This rule should be followed every time you answer an email or talk with a buyer on the phone or in person. As long as your intent is to sell meat or agriculture brood stock, you cannot be held in violation of the AWA if a buyer makes a pet of the rabbit after purchasing it.

Rabbits used for research in the United States are also regulated under the AWA, and a farm supplying them to research facilities must be AWA licensed and inspected. But believe me, research facilities will not even consider buying from a nonregistered rabbitry. They are far too closely scrutinized to purchase from a farm that is not fully AWA compliant.

Rabbits raised for research are also required to be purchased from AWA-compliant farms.

LAWS AND REGULATIONS FOR RABBIT MEAT

Under the Federal Meat Inspection Act, the USDA is responsible for the inspection of all swine, cattle, sheep, and goats processed in the United States. The branch of the USDA that is responsible for such inspection is the Food Safety and Inspection Service (FSIS). Under the Poultry Products Inspection Act, the FSIS inspects chickens, turkeys, ducks, geese, guineas, ratites, and squab. Rabbits do not fall under *either* act as of the writing of this book and thus are not mandated *federally* to be inspected—unless they are to be sold across state lines (including through the Internet). Then, the Food and Drug Administration (FDA) has jurisdiction instead of the USDA. The FDA will accept either a USDA stamp of wholesomeness following inspection by the FSIS or inspection of the carcass in an FDA-approved facility.

Regulations for rabbit meat sales within each individual state are different and are changing as the popularity of rabbit grows. Since rabbit is not federally mandated for inspection, each state has control over the regulation and inspection of rabbit meat for sale within that state. Some states allow the on-farm slaughter of a given number of animals. The rabbit raiser will usually have to be state licensed for meat packaging, but they can process and sell animals without any federal oversight. There will usually be regulations involving humane slaughter requirements, protocols for cleanup of facility and equipment, sanitary meat handling procedures, plans to ensure that meat packaging is done at proper temperatures, EPA-approved disposal of offal and wastewater, et cetera. A state-approved facility for on-farm processing may actually meet or exceed federal requirements.

Other states require that all rabbits are slaughtered at a federally inspected facility, and every carcass inspected to allow for sales within their state (even though it is not federally mandated)—meaning that all rabbits must have the appropriate USDA or FDA federal stamp affixed to the package as with any other mandatory-inspected animal species.

Still other states may fall in between and require that the animals be slaughtered and packaged in a USDA- or FDA-inspected facility but not require that each animal carcass be individually inspected (and no USDA or FDA stamp required on the label for sale within state lines). This last is important, as it means that even though farmers must use a federally inspected facility, they do not have to pay the federal inspectors a fee or be limited to the times inspectors are available. Thus, a slaughter plant can process rabbits on weekends when they cannot do other animals that require inspection. Many of these facilities would love to have their plant running on weekends!

Custom packing facilities, such as those licensed to cut up deer, are state inspected, but they are only able to slaughter or package meat intended for private consumption. These facilities cannot be used for rabbit meat (or any other meat) intended for resale to the general public.

Even if your state does not require any federal inspection, your customers (such as some restaurants) may still want it. You can have the FSIS do a voluntary inspection of your rabbits if you want to market to these establishments.

WHERE TO GET ASSISTANCE FOR COMPLIANCE WITH MEAT REGULATIONS

Since every state is different, contact *your* state to find out the exact hoops you must jump through to legally sell rabbit meat. For most states, this is under the Department of Agriculture, but others may have it under Consumer Services, Commerce and Regulatory Services, Food Safety and Compliance Programs, or State Food and Drug Divisions.

Your first act should be to contact your local Agriculture Extension Service officer for help. These folks are in all counties in the United States. The Cooperative Extension Service is a partnership between federal, state, and county governments to provide scientific knowledge and expertise to the farming public. Your local state agriculture college is often the point of contact for the agriculture extension services. In farming communities, they may have their own facilities.

All animals for sale across state lines must have a carcass-by-carcass federal inspection and a federal stamp of wholesomeness on the packaged meat.

When you are contacting either your extension service officers or your state's Department of Agriculture, remember that these people are not the enemy! On the contrary, they may be your best friends for trying to figure out what you need to do to legally "go commercial" with your rabbits. They don't write the laws; they are not trying to stand in your way; they are there to help you figure out how you can best meet the regulations. When you approach them, be ready with specific questions. Have at least a general notion of how many animals you would like to raise and where you might market them. The more directed you are, the better they will be able to help you. In addition to telling you about the slaughter and inspection regulations when done either "on-farm" or at a meat processor, they can tell you about such things as whether you need to hold a state retail food license and what is required for it in your state. In some cases, a fee paid to the state may be required. In other cases, they may want to do a yearly inspection of your rabbit freezers, have any meat scales state calibrated, and know how you plan to transport your meat safely.

The processor you plan to use for your rabbit, if you are not processing your own, may also have the answers to some of your questions. For example, they may weigh the rabbit at their plant when they package it and design the label for you.

This last might not sound like a big deal until you realize that there are specific requirements for each meat product label:

1. Product name (Domestic Rabbit Meat)
2. Inspection legend and packaging establishment number (the processor's name and address, and their USDA or FDA number if required)
3. Food handling statement (such as: Keep Frozen)
4. Net weight statement (not always required if you weigh at time of sale)
5. Ingredients statement (usually not needed for rabbit unless you are adding a marinade or other flavoring when packaging)
6. Address line (name and address of farm)
7. Nutrition facts (currently not needed by most farmers under the small business exemption, but this could always change)
8. Safe handling instructions (usually on a separate label for safe meat handling by the purchaser)

Any additional claim on a label, such as "raised without antibiotics" or "organic" will have to be preapproved by the FSIS and will require a written farm protocol and documentation.

There may be fewer regulations in your state for processing meat for pet food specially labeled "not for human consumption," but there are still controls. The reason for this is twofold: 1) no one wants

Our product, precut and vacuum-packaged, with all necessary label requirements, including a second label with safe handling instructions.

to see contaminated or spoiled meat fed to pets, and 2) some nefarious farmers have tried in the past to "get around" regulations for selling meat to the public by claiming they are selling it as pet food rather than human food. Contact your state's Agriculture Extension Specialists and Department of Agriculture to obtain the specific rules regarding legal pet food sales in your state if you plan to enter this market.

Don't use any "tricks" to get around the laws for rabbit meat sales. It is sad that many farmers feel so overwhelmed with red tape that they want to ignore it all. Yes, we have known farmers to sell "pet food" rabbit to folks (with its reduced processing regulations and controls), telling them that it is fine to consume it themselves if they want. Other farmers will claim on paper that they are selling live meat rabbits to people, and then offer a "free slaughter and freezing service" for them. They figure they sold a rabbit, not rabbit meat, so the regulations don't apply. These tricks, if you are caught, can result in heavy fines and possible prosecution. It will also invalidate any product liability insurance you might have (resulting in ruinous lawsuits if someone gets ill and claims it was by your product).

To protect yourself, be sure to document the dates, times, names, and titles of people you speak with regarding information you are given on selling rabbit in your state. Try to get information in writing when possible. Rabbit agriculture is a field where even some officials, who should know the correct answers, may not be as informed as they should be.

In short, you need to find the answers to the following questions.

- **Are there any *live* meat rabbit buyers in your area?** A meat processor? A "rabbit runner" who gathers rabbits from farms and transports them to a buyer? Do you have a local livestock market that will sell your live rabbits at auction? Do you have a farmer's market or flea market that allows the sale of live animals for meat (and a population of customers that will buy them)? Do you live in an area of the country where rabbit breeding stock sales might be a viable market?
- **What are the regulations for rabbit slaughter in order to sell *rabbit meat* to the public in your state through restaurants, farmer's markets, grocery stores, or as pet food?** Can you do your own slaughter on-farm with state oversight, or do you need a USDA/FDA-inspected processor? Does each animal carcass have to be individually inspected and a stamp affixed to the product, or can they just be packaged in an inspected facility without individual carcass inspection?
- **What are additional requirements for selling rabbit meat?** Do you need a retail food license? Inspection of freezers or refrigerators? What are the requirements for meat transport? What are the labeling requirements?

APPROACHING A PROCESSOR

If you cannot do on-farm processing and instead plan to approach a commercial meat packer, first check with your Agriculture Extension Service, the American Rabbit Breeders Association, or your state's Department of Agriculture to find any rabbit processors in your area that may buy your rabbits to slaughter and market themselves. If they are not buying live and marketing meat, there may still be a packing plant that is already processing rabbits routinely for other growers that will add you to their customer base for slaughter (but you will have to market the product yourself).

Any label claims must be FSIS approved.

If there are no packaging plants available at all, you may need to convince a processor that they should become licensed for slaughtering and packaging rabbit. In this case, you need to further scrutinize your retail market options so you can give your prospective processor an idea of how many animals you might be bringing per month.

Approach chicken processors to see if they might want to process rabbit. The equipment is much the same and rabbit is actually easier than chickens to process. If your state does not require carcass-by-carcass inspection of rabbit meat, the chicken processor may be able to slaughter and package rabbit on weekends when the inspectors are not available for their chickens.

There may also be specialty processors that your state officials may be able to steer you to that process less common animals, such as ducks, geese, quail, or pheasant, and that may want to expand to rabbit.

Do not overlook state-inspected (less commercial) Amish, Mennonite, or "Organic Only" processors in your area that may love to work with you. A few states even have traveling processing trucks that suit rabbit perfectly. The one thing you should keep in mind is that few processors will want to set up and take down their facility for 10 or 20 rabbits. You will need to have a market and breeders ready for a minimum of 40–50 at a time to make it worth their efforts.

Rabbits' small size, light meat, limited amount of blood, and lack of feathers make them extremely easy to slaughter and package. Vacuum-packaged rabbit has an exceptional freezer life, and "sell by" dates—if not required—should be avoided to prevent you from having to lower your prices for "older" product that is perfectly safe and still of very high quality. Of course, you will need to make sure that you have your freezers organized so you can rotate your product.

Rabbit being processed in a commercial packaging plant.

CHAPTER 9

MARKETS AND MARKETING

There are two categories of markets for rabbit: live rabbit and rabbit meat. Each requires a totally different approach to marketing. Live rabbit markets have two subcategories: selling live rabbits to be slaughtered and selling breeding stock. Packaged rabbit meat sales are 1) direct to the consumer locally or through the Internet, 2) through retail markets such as food stores or restaurants, or 3) as pet food. This chapter will discuss how to approach and market through each of these venues.

MAKING LIVE RABBIT SALES
PROCESSING PLANTS

One major outlet for selling live rabbits is to a meatpacking plant or rabbit processor. These outlets do all the marketing and distribution of the final product themselves nationwide. They will generally buy a certain number of rabbits per month at a contracted price per pound live weight. Most will buy all you can raise. The market for rabbit meat (at least at this time) is far outpacing the supply. Some will come to your farm to pick up the animals (if you are a major supplier); others will require you to transport them to the processing plant yourself or to a pick-up point where their "rabbit runners" will meet you and transport the rabbits from there. A major rabbit processor may have these pick-up points all over the state or region. (Or you may be unlucky and live in an area with *no* active rabbit processing plants and have to use an alternate marketing plan.)

SUPPORT
LOCAL
FARMERS

Becoming a contract grower for a processing plant is a very dependable year-round market for your fryers. (Note: these beautiful, heavy carcasses show excellent hindquarter muscling, are wide at the loin, and have a nice amount of fat over the shoulders.)

A word of caution: some of these buyers may try to get you to commit to selling them *all* of your crop, with no other outlets. You are much better off if you can contract to meet a certain set minimum number per month instead. If you sell all of your animals to one market outlet and that outlet folds, you are stuck with what may well be a major farming operation with no market outlet at all. Diversification is always optimal.

There is also the problem that if gas prices, feed prices, and utilities go up, and your rabbit buyer does not raise the price he will pay, you may be stuck with selling at a diminishing rate of return. If you have some outlets where you can set the price yourself, you can offset this. Unlike selling at a livestock sales barn, however, you at least have a guaranteed set price and can work to keep your costs down to make a profit within that framework (see Chapter 10 on making a profit).

These rabbit processors vary widely in what they are willing to pay. Some processors may only pay $1.00 per pound live weight and others closer to $3.00—it all depends on the size of their customer base and how many rabbit growers they have. That will obviously make a tremendous difference in your profit margin.

Some processors may take a fryer rabbit at 4.5 pounds, while others may require a minimum of 5.5 pounds. The difference in feed costs for this last pound will also affect your profit (remember, a rabbit's growth rate decreases as it gets older). Some processors will pay less for colored versus white rabbits or will not buy colored rabbits at all. They may require that the animals be under four months (or even under three months) to ensure tenderness. Most will require that they have never received antibiotics or other drugs (and will do random product testing to confirm this). On the upside, many of the processors are well aware of how much it actually costs to raise a rabbit and will try to pay enough to keep their rabbit growers in business—even adjusting the price they pay if feed costs rise dramatically.

SALE BARNS

Selling the majority of your crop at a livestock sale barn auction is the least desirable market venue, as there is virtually no control over the price you will receive. Though usually all of the animals will sell, they may sell very cheap. It all depends on the buyers that happen to be present on the day of the sale. You can easily lose money on a crop of fryers if a regular buyer doesn't show up that day. I would recommend only using sale barns for older cull rabbits or runts that do not meet your standards for other sales (and then only if you have enough such animals to cover the gas costs and auction fees and still realize a profit).

FARMER'S MARKETS—LIVE RABBIT SALES

Farmer's markets or flea markets where you are allowed to sell live rabbits direct to the public for use as food give you the option of setting your own price for your animals. You must be clear that you are *not* selling pets so that you do not run afoul of the Animal Welfare Act, as mentioned in the previous chapter.

Yes, business cards definitely increase your sales base.

The problem with this market is that many Americans no longer have the capacity to buy and slaughter their own food animals. Unlike chickens, rabbits are just too cute and will bring out the animal rights activists for sure. If you have an immigrant population near you more culturally adapted to slaughtering meat, you may want to invite them to purchase at your farm, thus saving the time, expense, and hassle of the market.

If you are considering this market outlet, remember that "rabbit fryers" are best (in our opinion) less than four months old and between 4.75 and 6.25 pounds, when they are the most tender and with a good meat-to-bone ratio. This is a limited time interval. If you can't sell your stock during that window, you are stuck with older, less tender "roaster rabbits," which sell for much less. Passing a roaster off as a fryer will ruin your reputation as an honest merchant. Even one cold, rainy weekend—when your crop is ready to sell but no customers want to brave the elements—can run you into a loss. You will also have to continue feeding the animals that don't sell unless you slaughter them all for your own food or take them to the sale barn.

There is little marketing that you will need to do for any of the above markets other than show up with your stock as agreed or on a predictable basis. We do suggest that you have business cards available. These can be printed very inexpensively and will almost always increase your sales. Your buyers need to be able to contact you via phone, website, or email to verify that you will be at a market or a rabbit pick-up point. Cards also help your customers share your information with other potential buyers.

BREEDING STOCK

Not every animal should be used for breeding, and you will still need an outlet for the rest of your crop. However, even selling just a few breeders per month can make the difference between a profitable month or not, due to the vast difference between what the market will pay for an

exceptional commercial breeder and a rabbit for dinner (see Chapter 10 on Setting Prices). At the time of this writing, breeding-quality New Zealand White does and bucks sell on average for $30–$40. If you collect the data we suggested in Chapter 6 for your records, you will have statistics that can lead to sales of "selected superior breeders" for closer to $50. You should be prepared to spend a good deal of time in aggressive marketing of breeding stock to achieve the best prices and a market for all your top animals.

If you are selling breeding stock as a significant portion of your business, you cannot *also* do live rabbit sales to the public via local livestock sales barns, farmer's markets, or on-farm live fryer sales. The reason is obvious: if people can obtain your rabbits cheaply at meat rabbit prices, they often will buy them in the hopes that they will be good breeders, even if you explain that you will not guarantee these animals like you will your brood stock animals. If people can buy your rabbits for $15–$20 as fryers, most will not want to pay $40+ for a selected breeder. Selling breeding stock works very well along with selling rabbit meat or selling live rabbits to a rabbit processor.

Believe it or not, people are willing to travel hundreds of miles for good commercial breeders. We constantly had a waiting list for breeding stock that stretched out for six months or more and ranged from 1 to 100 animals per buyer. Appendix A shows an example of a rabbit purchase agreement. We suggest a similar written agreement be utilized to prevent having to hold breeders too long.

To sell breeding stock, your facility must be kept clean and disease free.

To achieve this level of sales requires two things: 1) quality breeders that you can stand behind and 2) willingness to market via the Internet. Local breeding stock sales will eventually be limited in numbers, and you must cast a wider net to make sure that all of your potential brood stock have buyers. (We will discuss more on website design later in this chapter.)

The next thing to consider is time. Selling breeding stock requires time on the phone or email answering questions for potential buyers regarding your selection criteria, rabbit raising in general, and raising rabbits for commercial sale. The better constructed your website, the fewer questions you will have to answer. You will also have to be present when buyers come to tour your facility or buy stock, and you can count on an hour or so answering additional questions.

If you absolutely hate sales or talking to people, or you don't have this time to invest, you should reconsider this as a market outlet. Unlike folks just buying a rabbit or two to eat, if a buyer is considering buying 10 to 20 rabbit breeders for hundreds of dollars, they will want to know a lot about your operation! You will also need to keep your place clean at all times and your rabbits disease free. You must also give up a bit of privacy. But if you like people and enjoy "talking rabbits," and you are proud of your facility and your stock, this can actually be the most rewarding part of rabbit sales—both financially and personally.

It may take a bit for you to select your rabbits and establish your reputation as a breeder. However, if you start with good, solid stock, considering the fast generational turnover in rabbit breeding, you can begin to offer breeders within a year or two. If you are a large enough establishment to be able to offer more than one breed, you will increase the chances that people will travel a fair distance to your facility.

We were located "out in the boonies" like many farms, but in Middle Tennessee, we were central in a larger respect to many areas of the country. We had buyers travel from as far as Montana, Texas, Florida, Wisconsin, and New York for our breeders. If you are geographically located in southern Florida, northern Maine, or western Oregon, you will be somewhat more limited in potential buyers.

If you have a smaller farm and want to just sell your breeders locally, you will have to work to keep your market viable. This means advertisement. You can advertise free on bulletin boards at feed stores, farm equipment stores, country food stores, and anyplace where prospective raisers might congregate. You can also take out fairly cheap advertisements in small local papers, farm bureau magazines, or local classified-sale papers. If this doesn't bring in enough business, you can set up a booth on occasion at a farmer's or flea market to talk to people (with business cards and flyers for your farm).

Some feed stores (if they carry rabbit feed and equipment) may let you set up a table for a day outside their door to talk to people. You can contact your County Agriculture Extension Agents and see if you can give a seminar on rabbits and the rabbit industry. Many states run websites encouraging the "buy local" movement and will put your farm information on their site for free. Find out if there are any local prepper, survivalist, or homesteader groups in your area who are interested in a self-sustaining meat source. (These were some of our best customers.)

MAKING RABBIT MEAT SALES

If you have waded through the necessary regulations outlined in the previous chapter and found a state-approved slaughter facility or a USDA/FDA-certified processor or set up an approved processing facility on-farm, the possibilities are almost unlimited for marketing rabbit meat to the public. Rabbit meat can be sold at farmer's markets, restaurants, food stores, and pet stores, as well as through on-farm or Internet sales. We constantly had more market outlets than rabbits in the freezers! Below are the pros and cons of each of these possible outlets and some ideas on the best way to approach them.

FARMER'S MARKETS—MEAT SALES

Farmer's markets are a lot of fun but can be time consuming, and you need to be selective to obtain the best results for your effort. With the current appetite (pun intended) for farm-to-table foods, there will

Spend some time researching your local farmer's market options.

usually be several markets within driving distance of your farm. If there is more than one, see if any of them conduct promotions or events to attract customers to their venue.

Of course, markets attached to larger cities will have more foot traffic than ones in small communities. Having a combined flea market/farmer's market may attract more traffic, but many of the shoppers may not be looking for food. If it is an all-food market, you know that everyone you see is looking for good stuff to eat. A market that is local artisans and farmers is often a good combination, as the "buy local" trend is in your favor.

You want to avoid markets that are, in reality, just wholesalers (not really farmers). You can tell this by simply looking at the products offered for sale. If you see people selling pineapples and oranges in Pennsylvania or "fresh-picked" apples in March, you can be pretty sure these are *not* farmers.

Customers who shop at farmer's markets generally fall into two categories:

1. **Those looking for healthier local options.** They don't want any preservatives, insecticides, added hormones, or antibiotics. They want to support local producers and are willing to pay more to "their" farmer. These people want to be educated about their food and will resent "farmer's markets" made up of non-farmer wholesalers. These customers make excellent converts to rabbit meat when they discover how beneficial it is. They are quite willing to pay more for healthy options.

2. **Those looking for a good deal on cheap food.** They will frequent the wholesale-type market rather than organic or locally grown markets. They may balk at rabbit prices at first, but—with good salesmanship on your part—they can still become customers when you explain that an $18–$24 large whole rabbit can easily feed a family of four (at $6 or less per meal).

Many immigrant populations or foreign nationals also visit farmer's markets looking for a shopping experience that may be more like that in their home country. They are a demographic you should explore, as many will be willing to pay well for a taste of "back home."

When you are deciding on your best market venue, check out the other meat sellers. They will, on the one hand, be your competition for the "meat dollars" in the market, but on the other hand, an organic beef producer or someone selling pasture-raised lamb (which we also did) will already be attracting high-end consumers looking for quality meats. Markets that have meat sellers, cottage food industries, and local craftsmen may operate all year long, not just in the produce-growing season, allowing more dependable year-round sales.

Another consideration when choosing a market is its requirements for attendance on the part of the farmer. Some will only allow sellers who attend every weekend, while others will be willing to schedule certain weekends for you. This allows you time off for other projects, or even to attend two different markets in diverse areas. If the market has a website or social media presence, keep yourself in the "news" and have the weekends you will be attending the market published so your customers will know when you will be there.

Some of the larger markets may have events ranging from cooking shows (where you can cut up and barbecue a rabbit), runs or walks for charity (which bring out those "heart health conscious" consumers), informative talks on sustainable agriculture (in which rabbit raising fits perfectly), or local music talent performing to draw customers.

Once you have chosen your market, you need to consider the cost of the booth and what is provided, such as electricity (if you are bringing small freezers), tables, or shade. Next, you should inquire if product liability insurance is required. Even if it is not, it is a good idea to have it anyway. We had an umbrella policy in which we had the insurance company specifically write in a product liability clause. You definitely do not want to ever find yourself in a situation where you lose your farm over a rabbit sale!

Once you have chosen your market, you are ready to present your rabbit to the public. In

A market display does not have to be expensive to be effective.

Farmer's market patrons want to be educated by "their" farmers.

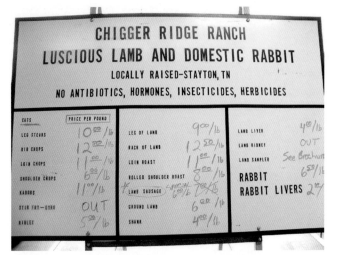

CHIGGER RIDGE RANCH
LUSCIOUS LAMB AND DOMESTIC RABBIT
LOCALLY RAISED—STAYTON, TN
NO ANTIBIOTICS, HORMONES, INSECTICIDES, HERBICIDES

CUTS	PRICE PER POUND				
LEG STEAKS	10⁰⁰/lb	LEG OF LAMB	9⁰⁰/lb	LAMB LIVER	4⁰⁰/lb
RIB CHOPS	12⁵⁰/lb	RACK OF LAMB	12⁵⁰/lb	LAMB KIDNEY	OUT
LOIN CHOPS	11⁰⁰/lb	LOIN ROAST	11⁰⁰/lb	LAMB SAMPLER	See Brochure
SHOULDER CHOPS	6⁰⁰/lb	ROLLED SHOULDER ROAST	8⁰⁰/lb	RABBIT	6⁵⁰/lb
KABOBS	11⁰⁰/lb	LAMB SAUSAGE SPECIAL	6⁰⁰/lb	RABBIT LIVERS	2⁰⁰
STIR FRY—GYRO	OUT	GROUND LAMB	6⁰⁰/lb		
RIBLET	5⁵⁰/lb	SHANK	4⁰⁰/lb		

Add buzzwords such as "locally raised" or "antibiotic free" to increase direct sales appeal.

this setting, you are trying to create a relationship with your customers as part of the "farmer's market experience." To the left is a photo of a simple whiteboard display. You can order more professional signs as you progress.

Appendix B depicts a copy of a brochure that we utilized to teach people about us as rabbit ranchers. You should include in your brochure how the animals are raised, where your meat can be purchased, your contact information, prices (unless they will be fluctuating), where it is packaged, and how to cook it. Information about the advantages of rabbit meat to personal health and to the world as part of sustainable agriculture can be highlighted as well. We carried a couple of specialty cookbooks to add to our farmer's market display appeal. Alternatively, you can print recipe cards for tasty rabbit dishes—with a new one every month to encourage sales.

Your presentation should include catchwords like "locally grown," "all-natural," and "gourmet-quality" and information such as "raised without antibiotics, insecticides, herbicides, or added hormones" (if this is all true, of course). Prices should be clearly marked, especially if your market is a busy one—you do not want people to wait in line just to ask how much a rabbit costs. Your website or contact information and business cards should be prominent in your market booth.

If your processor does not weigh the rabbit and add that to the label, you will need to have a state-certified scale at your booth. Remember to bring plenty of change, a way to accept credit cards, a receipt book, and your state food handler's license every time you attend.

You *must* be willing to talk to people about your product and your farm to be an effective salesman. This is the fun part, and we completely enjoy teaching people about a new and healthy food source. Yes, you may have to fend off the occasional animal rights individual. These can usually be handled by explaining sustainable agriculture. Talk about the rabbit's spot in the food chain and about humans being naturally omnivorous. Give them a farm brochure. Tell them you can respect their belief but feel you are humanely and compassionately providing food for many. We found that when we passionately and sincerely explained our views, we *never* had a problem.

Of course, do not *ever* put cute, fuzzy rabbits on your business cards, brochures, market displays, etc. No, it is *not* a good idea to sell stuffed-animal toy bunnies from your booth! You are not trying to hide the fact that rabbit meat comes from rabbits, but many Americans are no longer at ease with connecting their packaged meat with the live animal.

> **You *must* be willing to talk to people about your product and your farm to be an effective salesman.**

When dealing with animal rights activists, be passionate and sincere. Do not apologize for being a farmer, but be respectful of their beliefs and calmly explain your views. While we may not agree on many points, they have been instrumental in rectifying problems in several animal industries.

A rabbit pet owner may look at you as if you are slaughtering dogs and cats. Do not try to convince them they cannot be pets. Agree with them that a few rabbits do have a cool personality, and so do many goats and cows. But unless a person is going to be 100 percent vegan, he or she is being hypocritical by singling out rabbits as unsuitable for food. They have been food for thousands of years for many species of animals around the planet, including humans. Transition your conversation to your dedication to humane farming with the physical, social, and psychological needs of your animals met to the best of your ability.

RESTAURANTS

For some reason, many farmers are intimidated by restaurants. If you are following all the regulations for meat sales in your state, then restaurants can be a *very* lucrative and dependable outlet for your fryers. First, you need to decide which to approach in your area. You will waste your time trying to sell to chain-type restaurants, where the menu is formulated at corporate headquarters and is the same throughout the United States. Look for individually owned restaurants that are trying to compete for the "dining out dollar."

The most obvious choice is the farm-to-table venues, where they are already educating their customers about locally grown produce and healthy food choices. They may be interested in your farm brochure to promote their rabbit dishes, or you may want to write a brochure dedicated more specifically to restaurants—with facts about humane animal agriculture, the low fat and high protein qualities of rabbit, and the importance of supporting local farms and sustainable agriculture.

If you do not have a farm-to-table establishment in your area, do not despair. A second, extremely viable choice is the high-end gourmet restaurant. The chef in such a business may be very excited to try a new meat. Rabbit does very well with many cooking styles and with a wide range of sauces and spices. These upscale enterprises also generally have more expensive menus than a "mom and pop" type place and are more able to afford to offer rabbit.

Some international restaurants may be interested in rabbit as a dish that they can promote to their customers—either as something their traditional clientele might be familiar with or something their American customers would like to experience.

If, like many farmers, you are in an area with more "down-home" type dining choices, there are still options. For example, if you have a couple of buffet-style establishments that are competing for that Sunday afternoon brunch crowd, one may buy rabbit for a weekly special on their buffet to attract more customers their way.

Rabbit does real well on the barbecue, and barbecue connoisseurs are more than willing to buy it. With pork and chicken being popular and much cheaper alternatives, this venue will depend on the local economic demographics. Barbecue restaurants located in higher-income areas may do quite well with rabbit.

Once you have chosen a restaurant to approach, there are a few things to keep in mind to have a successful sale. First, who decides on the menu? Is it the chef, the owner, or the manager? Although the chef should almost always be consulted and included in correspondence, others may make the final decisions. (Interesting the chef in rabbit is key to restaurant sales—but actually offering recipes may be

You must be dependable and loyal to maintain restaurants as customers.

Stressing the versatility of rabbit is often key to making restaurant sales.

insulting!) If a chef has absolutely no interest in experimenting with rabbit, you can pretty much write off that restaurant as long as that chef is there.

Appendix C is an example of a flyer we left at restaurants or emailed to them. Be professional; never use a restaurant's web page that is intended for dining reservations or comments on their food to make an offer of rabbit. If you cannot obtain an appropriate Internet contact point, you may need to visit in person.

In-person visits to prospective restaurants also require planning. Make an appointment to talk with the owner, chef, or manager when possible. Always choose a downtime to visit—not during or just before the lunch or dinner rush! Go eat at the establishment. Talk to your server and explain that you are interested in offering a new menu item of rabbit. (A friendly server can be your most valuable ally!) Get as much information as you can regarding to whom you should be talking, how often the menu changes, etc. Leave a business card, brochure, or restaurant flyer (or all three).

The delivery door to the kitchen, where the chef will be located, is another place you should go. Do not disturb a busy chef, but leave information for him or her. Chefs will appreciate being consulted. If the chef or owner seems interested, have a frozen rabbit with you to give as a free sample. Many people have difficulty even picturing what a whole rabbit looks like! We had more than one restaurant ask us to come in to show the best way to cut up a thawed rabbit for cooking.

Understand that a restaurant will usually publish a menu in advance, and its customers will be (very) upset if a meal they desire is not available. A restaurant never wants to say they are out of something! Try to make the chef and/or owner understand that you work to breed a certain number of rabbits three to four months in advance. If a major increase in sales is anticipated (for example, in February due to Valentine's Day), you need to be kept in the loop as much as possible.

When starting with a new restaurant, word may spread quickly that its rabbit is out-of-this-world yummy, leading to a significant and unanticipated upsurge in orders. Since rabbit (in vacuum packaging) has a long freezer life, always have backup rabbit ready to go. Do not try to start up with too many restaurants at once until they—and you—are able to predict the sale volume.

To sell at food stores, a precut rabbit with distinctive labeling is most important, as this will be what draws customers to your product.

Loyalty to your restaurants is a very important aspect of your reputation. If you have a reliable customer, you do not want to approach its main competition to offer rabbit. If the competition asks you to supply it with rabbit, you may have to inform the second restaurant that you can only raise so many rabbits per month and do not feel you can reliably serve them at this time. Do not ever close the door completely, as your first customer may change hands, change chefs, or fail completely.

There will be more discussion on setting your prices in Chapter 10; however, it is good if you can offer your restaurants a "deal" as much as possible. The majority of their meat is bought at wholesale prices, not retail, and they are used to that break. If they are buying a large number at one time and on a set schedule, you can actually save enough money on gas delivery and marketing costs to allow you to give them at least a bit off and still make a tidy profit.

FOOD STORES

Most of the considerations mentioned above for picking and contacting a restaurant are the same as those to examine when you are choosing and approaching a food store. Look for the appropriate establishment first—the ones that focus on local products, healthy options, and international foods are obvious choices. However, larger chain stores may be willing to sell your rabbit, as many of these supermarkets, unlike chain restaurants, have some discretion over the products they stock.

Freezer space is often at a premium in food stores. Offer to check on the stock and bring replacements on an as-needed basis so that they can dedicate less room to you, at least at first—until it is a proven product and customer draw. A nice label and attractive packaging will help this form of marketing more than anything else.

It is possible that only cut-up rabbit will sell well in this venue, as many people are intimidated by trying to cut an unfamiliar carcass. Requiring your processor to cut up the rabbit will increase their charges to you (which you then have to pass on to the food store customer). The main competition you will have at food stores is the cheap, imported, foreign-raised rabbit. If a store is already carrying this, it may be unwilling to change to a more expensive alternative. However, if it starts with a meaty American rabbit, with optimal meat-to-bone ratio and attractive vacuum packaging, it will have difficulty switching to a less desirable product.

PET FOOD SALES

Pet food sales can be a major fryer outlet or just a sideline market for excess pinkies, runt fryers, and older rabbits. Some pet food companies want the whole rabbit, with skin, hair, and head left on and no removal of internal organs. That is, the rabbit is euthanized and just put whole into the freezer in a bag with no packaging. We constantly had calls from such companies wanting to buy our rabbit. Whole, intact, frozen rabbit is also common for reptile food (for snakes, alligators, and such). Local zoos might buy rabbits to provide the occasional treat to their large carnivores.

Pet food can also be processed similarly to food for human consumption, with the head, skin, and internal organs all removed, and the carcass packaged whole and labeled as pet food. It is generally sold like this direct to the public on-farm, at farmer's markets, or through veterinary clinics and pet stores to those people wanting to feed their dogs or cats raw meat (we got a lot of these requests). As mentioned in the previous chapter, just because it is "not for human consumption," there are still slaughter and packaging regulations that must be followed.

Certain internal organs, such as liver, kidney, and hearts, can be packaged separately and are prized for pet food for everything from cats and dogs to some kinds of lizards.

INTERNET-BASED MEAT SALES

Meat sold across state lines must be packaged in a USDA- or FDA-inspected facility, and the carcass inspected with the proper stamp

Rabbit meat sales to pet food companies can be a major market outlet or cull rabbits and rabbit organs can be sold locally as a sideline.

of wholesomeness on the label (see Chapter 8). Reasonably priced rabbit sold via the Internet will almost always have far more market than a single grower can supply. It is an excellent venue for a larger operation or a group of growers acting as a cooperative.

Remember that you must add in the costs of shipping, shipping containers, and dry ice hazardous material charges, and you will need a good deal of time for packaging orders. You obviously need a website and some sort of a Paypal vendor account or a security system that will allow you to engage in safe e-commerce. In fact, credit card companies will insist that you comply with certain security protocols to be able to accept their cards. There will be transaction fees charged to you for credit card payments.

NOTES ON WEBSITE MARKETING

If you are planning to make more than just a few dollars on your rabbit enterprise, you will need a website to, *at the very least,* promote your breeding stock or local meat sales, even if you do not ship or sell meat across state lines. This is a requirement today in any business environment, and as younger people become future consumers, it will become an increased necessity (along with social media-based advertising).

Before we go further, we should mention about web-hosting companies. A website host is a place (known as a server) where your website is located so that people surfing can find it. To try to host a website on your home computer, you need fast 24/7 connectivity, server software, a very stable and reliable operating system, and strong security.

Even if you have all of the above, your Internet service provider may not allow you to host a website because your IP address (the unique address of your computer on the Internet) changes every time you connect (this is called a *dynamic* IP address). To successfully host a website, your computer must have a *static* IP address so that people can always find and link to it. Your Internet service provider can assign your computer a static IP address but will usually charge you additional money to do so. Many companies specializing in web hosting charge very little to give you your own domain name associated with a static IP address on their secured server. They may add filtered business email, e-commerce tools, and backups for your website. These deals can start as low as $5-$10 per month and go up from there, depending on the level of service you require.

There are also website design companies (as opposed to website hosts). These companies charge you money for initially designing and for updating your website. If you elect to take this route, you will still have to do a good deal of work. They may know website design, but it is practically guaranteed they will not know enough about rabbits (much less *your* rabbits) to be of much help in writing the copy for you. A complete course in website design is far beyond the scope of this book, but we will give you a few things to think about below.

There are a number of free website design programs available if you do not want to pay a professional designer. Many of these programs are referred to as "drag and drop," meaning you can drag photos or graphics into an empty box on a pre-programmed template, and the program will automatically place it there. There will be other boxes on the website dedicated to text that you can type in yourself. You do not have to know any computer programming language at all! You can also purchase website design programs that allow you more freedom than using a set template but are similar to use. These usually have easy-to-follow instructions or technical help available.

When Internet marketing, identify your target audience and provide *relevant* information.

When you are thinking about the text to use in your site, ponder what you are selling and to whom, and try to figure out what *they* would type into a search engine (such as Google, Bing, Yahoo, or Dogpile) to find you. This will give you the keywords you need to work into your text. Search engine result ranking (whether your site shows up at spot number 10 on the result list after a query or way down at number 100) is determined by the content of your website as the search engines see it. The search engine will look at your site to determine what it is about before it adds it to the massive database it maintains.

Search engine companies seek to provide the searcher with the most *relevant* sites to answer their inquiries (or else people will stop using their search engine). They do not give a hoot whether or not *you* sell anything! Their responsibility is to the person querying the web, not to the person posting a website (unless you are paying the search engine company an advertisement fee to put you at the top of the list).

This means that you need to include plenty of words like "rabbit meat for sale," "raising rabbits," "rabbit breeding stock," "rabbit brood stock," "New Zealand White rabbits," "Californian rabbits," "rabbitry," "commercial rabbits," etc., in several places in the text. You need to go further than this, however. You want to include "sustainable agriculture," "heart-healthy foods," "organic gardening," "urban farming," "alternative farming," and "locally grown produce" to get more people visiting your site. Do not forget to include your farm name and location in the text.

They say a photo is worth a thousand words, but this is *not* the case when you are talking about website search engines. The search engines only look at text, not images or movies. Photos and graphics will improve your website's appeal and impact but not do anything to get customers *to* the website.

Mouthwatering photos can be persuasive marketing tools.

Also consider the titles you give the different pages of your website. These are looked at by most search engines and need to be carefully constructed to describe the page accurately and help the search engines precisely classify you. For example, you do not want to use "David's Website," "My page 1," or "Meat" for your page tags. Instead, use phrases such as "Chigger Ridge Ranch Rabbits Home Page," "Tennessee Rabbit Meat for Sale," "New Zealand White Rabbit Breeding Stock," "Rabbits as Part of Sustainable Agriculture," and "Heart Healthy Rabbit Meat" to title your pages.

Many website design programs allow you to add things called meta tags. These are basically a list of keywords like the ones you work into your text that describe your site for the search engine. Not every search engine uses them, but never ignore the option if it is available to you. List *any and every* keyword you can think of that someone might use in a search!

The more links to and from your site, the higher the search engine will rank you. Talk to any restaurants, stores, or farmer's markets you use and ask if you can link to them and if they will link back. Link to local rabbit clubs or rabbit breeder associations if they will allow it. There are many state-sponsored sites that have lists of locally raised farm produce and will be happy to create links to your website. Even if you cannot get a reciprocal link, the more the search engines see you as a "hub" where people go, the higher your site will start to rank when people do a search.

Finally, have patience with the search engines. They take a while to index and rank you, and you may not see yourself up toward the top of the list until you start to have traffic to your site.

The actual content of the text in your website will vary with your own farm, of course. Always put yourself in the shoes of the customers. What information would *you* want if you were them? With the advent of social media, it seems many people have developed the impression that everybody on the web is somehow interested in every aspect of their lives. This is not how you should be thinking if you are trying to market a product. You are not doing Facebook with your friends! Non-farmers may like a few personal farm tales, but in general, you should be professional and concise and provide relevant information.

If you are selling breeding stock and your prospective buyers have never raised rabbits, they might want to know the answers to basic questions like:

1. How hard is it to raise rabbits?
2. Why is rabbit meat a good option for a family?
3. What do I need to feed a rabbit to keep it healthy?
4. What equipment should I have before I buy my rabbits?
5. How many rabbits can I raise in a year from selected breeders?
6. How many rabbits do I need to feed a family of four?

If the reader is a commercial rabbit raiser already, he or she will be looking for:

1. How you select your stock
2. Statistics on your farm's productivity
3. How long you have been rabbit ranching
4. How many does and bucks you keep
5. How long your waiting list is for breeders

If you are selling rabbit meat, the buyer wants to know totally different things, such as:

1. How they are raised
2. How they are packaged
3. How many people one rabbit will feed
4. How to cook them
5. How much they cost per pound
6. Where your product can be purchased
7. Why rabbit meat is so great

Think of all possible audiences and address as many as you can. Visiting your website will be preppers, back-to-nature types, commercial growers, urban farmers, health-conscious consumers, local-market advocates, sustainable agriculture fans, foodies interested in exotic meats, and international cuisine connoisseurs. With all of this, you should not lack for things to say!

CHAPTER 10

COSTS
AND PROFITS

VARYING MARKETS TO AFFECT PROFIT MARGINS

If you want to make money raising rabbits, you should:

- Keep production records
- Weigh and measure them
- Select your breeders accordingly
- Pay attention to sanitation, health concerns, and breeding schedules
- Provide adequate feed
- Market your product (don't just dump them at a livestock market)

Rabbits are so efficient and prolific, it is hard *not* to make money with them! This chapter is designed to help you determine your costs and set your rabbit price to keep you in the black and realize a profit.

Business An

> **Rabbits are so efficient and prolific, it is hard *not* to make money with them!**

The first objective is an accurate record of monthly expenses related to rabbits. Since rabbits have a short turnover rate—that is, only 2–4 months from birth to having a crop of rabbits ready to sell—we recommend that you keep records monthly rather than seasonally or yearly as other agricultural industries can. If you stagger breeding, you usually have a rabbit crop to sell every month, unless you are a very small operation.

The following eight cost categories are also important to have recorded (and *every* expense documented with receipts) for your taxes at the end of the year. Note that we have not included start-up costs here, just the normal monthly expenses. Start-up will be addressed later in this chapter.

1. **Feed** costs should not be divided into what it takes to raise the crop of fryers and what it costs to feed the breeders and replacement rabbits you are raising for yourself. In performing these calculations, ultimately you want to be able to set your rabbit price so the fryers you sell are making enough to feed themselves while they grow, but also support all the other rabbits. So your feed costs listed should be all of your feed for that month. Note that the feed costs are higher than all of the other costs combined. This is usually the case, and is why so much of this book is dedicated to such things as the rabbit digestive system and selecting your rabbits for heavy milking does, feed efficiency, and fast kit growth.

2. **Bedding** is simply the nesting material you use each month.

3. **Utilities** can be difficult to determine if your rabbit utilities are included in with your household bill, but you should be able to estimate your percentage of "rabbit-derived utilities" by comparing the monthly bill after you started raising rabbits with the same month's bill from the previous year before you started raising rabbits.

4. **Veterinary** costs for a rabbit enterprise are usually low. A sick rabbit rarely lives long, and veterinary trips are generally not cost-effective unless you are obtaining help diagnosing a problem that might affect other rabbits. You might purchase such items as ear mite medication or topical antibiotics which should be listed under this category.

5. **Equipment** costs here should be small-item monthly expenditures (not larger start-up equipment). It should include *everything,* though! You will be surprised at how many small farm-related things you buy in a month. These items will vary a lot, but remember that you are also tracking costs for your tax deductions, and the more you document each month, the more you will be able to deduct. Include animal equipment, such as replacement feeders, but also list cleaning supplies, office materials, and even your new work gloves.

> **100 kits per month can be raised by one person or a couple working other jobs. More rabbits than this may require hired help and the associated costs of labor.**

It does not take magic to make a profit with rabbits. They are unquestionably one of the more controllable and predictable agriculture enterprises.

6. **Marketing** costs are any advertising or sale-related costs: website hosting, farmer's market or sale barn fees, printing brochures, and advertisements. These costs may be eliminated if you do not sell anything directly to the public.

7. **Slaughter/processing** costs are only included if you sell rabbit meat. If your market is 100 percent live rabbit sales, this cost is eliminated.

8. **Gas** costs can vary widely. If you live 35 miles from your feed store, 100 miles one-way from your slaughterhouse, and you deliver rabbits to restaurants 45 miles away, obviously your gas costs are going to be a far more significant portion of your operating costs than a farmer with a rabbit buyer who picks up live rabbits monthly on his farm and only counts a few trips to the feed store in his gas costs.

The costs presented here should be viewed as examples, not costs you should expect for your particular operation. They are "real life" in that they are *similar* to what you might see in raising a hundred kits a month—but every operation is different.

Why are we showing 100 fryers per month as an example for this analysis? One reason is that 100 rabbits per month is a reasonable number to be raised by a single person or a couple working outside jobs with rabbits as a sideline business. If you work full-time on a farm, you will be able to increase the number of rabbits you can raise per month. The daily chores on a rabbit farm do not take long, but the weekly cleaning, breeding, weaning, weighing, and trips to feed stores, processors, and market outlets can be time-consuming and make it difficult to work around other jobs.

The first two tables show the costs and income for an example farm that we will call Farm A, which has diverse marketing outlets—breeding stock, sales to a contract meat rabbit buyer, and direct sales of rabbit meat to the public at farmer's markets, restaurants, and stores. Additionally, because of exposure at farmer's markets, this farm was able to sell some rabbit manure to the organic farmers, culled roaster rabbits as pet food, and even rabbit livers to a local chef.

In the Farm A example, the costs are $865.75, as shown in Cost Table Farm A, and the income table for Farm A shows $2,085.20 generated per month. This enterprise has made a profit of $1,219.45 for that one month ($2,085.20 − $865.75 = $1,219.45). We will show you several different business plan examples in this chapter—all but *one* of which make a profit.

Cost Table Farm A (100 kits per month)

EXPENSES	DESCRIPTION	COST
Feed	pellets, hay, supplements	$500.00
Bedding	straw, nest inserts	$35.00
Utilities	rabbit barn electric and water	$60.00
Veterinary	ear mite medication	$7.00
Equipment	2 feeders replaced, marking spray, wire brush, bleach	$35.00
Marketing	website hosting, printed brochures, farmer's market booth	$55.00
Slaughter/processing	45 rabbits @ $2.75 per rabbit (40 fryers and 5 roasters)	$123.75
Gas costs	400 miles with 20 miles per gallon = 20 gallons @ $2.50 per gallon	$50.00
	TOTAL COSTS FOR JANUARY	**$865.75**

The first five cost categories are generally fixed for 100 kits, unless you change your genetics, barn type, or husbandry protocols. The last three categories are variable, depending exclusively on how the rabbits are marketed. For example, the gas costs are highest on this farm (compared to Farms B–E) due to the diverse marketing outlets requiring deliveries (seen in Income Table Farm A).

Income Table Farm A

WHERE RABBITS SOLD	NUMBER OF RABBITS OR PRODUCT SOLD	PRICE PER RABBIT OR PER POUND	INCOME
Breeding stock	20	$30 per rabbit	$600.00
Live rabbit buyer	40	5.75 pounds average, 230 pounds live weight sold @ $2.60 per pound	$598.00
Farmer's market	20	2.8 pounds average, 56 pounds total sold @ $6.95 per pound	$389.20
Restaurant/store	20	2.8 pounds average, 56 pounds total sold @ $6.25 per pound	$350.00
Rabbit manure and worms	1 load manure	$40 per load	$40.00
5 roaster rabbits packaged for pet food	5 large cull rabbits sold for pet food	4.6 pounds average, 23 pounds total sold @ $4 per pound.	$92.00
Rabbit livers	4 pounds	@ $4 per pound	$16.00
TOTAL INCOME FOR JANUARY			$2,085.20

With a very diverse marketing scheme, this farm has a year-round stable income, and more than half of the sales are able to be priced by the farmer (with only 40 rabbits sold at contract prices to a live rabbit buyer). Sales of secondary products are also a source of income (manure, cull rabbits, and rabbit livers). With such varied markets, this farm could raise over 100 rabbits and still expect to be able to move them all.

> **Feed costs are *the* single highest expense on a rabbit farm (often more than all the other costs combined) and the most important cost to contain.**

Farmer B does not have the time to do any marketing but found a contract rabbit buyer who will come to the farm and purchase 100 live kits per month. The cost table will be completely different, as this farmer will eliminate marketing and slaughter costs. His gas expenditure is also down, with the only travel being to the feed store. With the trimming of these expenses, the costs are decreased to $662.00 (Cost Table Farm B).

Cost Table Farm B (100 kits per month)

EXPENSES	DESCRIPTION	COST
Feed	pellets, hay, supplements	$500.00
Bedding	straw, nest inserts	$35.00
Utilities	rabbit barn electric and water	$60.00
Veterinary	ear mite medication	$7.00
Equipment	2 feeders replaced, marking spray, wire brush, bleach	$35.00
Marketing	none	$0.00
Slaughter/processing	none	$0.00
Gas costs	200 miles with 20 miles per gallon = 10 gallons @ $2.50 per gallon	$25.00
	TOTAL COSTS FOR JANUARY	$662.00

The fixed costs are unchanged, but the variable costs are lower on this farm due to the one-source market (a contract buyer). There are no marketing or slaughter costs, and gas expenditures are reduced.

Income Table Farm B

WHERE RABBITS SOLD	NUMBER OF RABBITS OR PRODUCT SOLD	PRICE PER RABBIT OR PER POUND	INCOME
Live rabbit buyer	100	5.75 pounds average, 575 pounds @ $2.60 per pound	$1,495.00
		TOTAL INCOME FOR JANUARY	$1,495.00

This simple marketing strategy, with all rabbits sold to a contract buyer at a set price, gives predictable income for this farm. But if fixed costs such as feed prices begin to rise, the farmer cannot raise his prices to compensate!

However, the *income* for Farm B is also altered, in that roaster rabbits, rabbit livers, and manure are not sold, due to the decrease in customer contacts. With the rabbit buyer contracting at $2.60/pound, the income for Farm B is $1,495.00 and the profit is $833 dollars per month ($1,495 − $662 = $833). This is less than Farm A's $1,219.45 profit, but still not bad for the reduced effort involved.

At Farm C, there are no local restaurants or stores that want to carry rabbit. There are no live rabbit contract buyers. However, there is a very active farmer's market, and Farmer C is also interested in selling some breeding stock. With this business plan, the marketing costs are increased, with more frequent trips to the farmer's market and thus higher booth costs. The slaughtering/processing costs are also increased, with 80 percent of the kits sold direct to the public as meat.

Costs are up to $1,003.25, but since the farmer can set the price for all of his product, he can cover these added costs. He has also made contacts to sell manure, older roaster rabbits, and rabbit livers. The income from this farm is $2,304.80 per month. The profit is $1,301.55 ($2,304.80 − $1,003.25 = $1,301.55). This is $82.10 more than Farm A and $468.55 more than Farm B.

Though the profit is the highest at Farm C (with all retail meat sales direct to the public at a good price utilizing farmer's markets), there is the problem that this farm must sell a *lot* of rabbits year round without any set presold numbers per month. If the winters have decreased traffic at the market, Farm C may have to increase its sales in the other seasons to move all their product. Also, if they decide to raise

Cost Table Farm C (100 kits per month)

EXPENSES	DESCRIPTION	COST
Feed	pellets, hay, supplements	$500.00
Bedding	straw, nest inserts	$35.00
Utilities	rabbit barn electric and water	$60.00
Veterinary	ear mite medication	$7.00
Equipment	2 feeders replaced, marking spray, wire brush, bleach	$35.00
Marketing	website hosting, printed brochures, farmer's market booth	$95.00
Slaughter/ processing	85 rabbits @ $2.75 per rabbit (80 fryers and 5 roasters)	$233.75
Gas costs	300 miles with 20 miles per gallon = 15 gallons @ $2.50 per gallon	$37.50
	TOTAL COSTS FOR JANUARY	**$1,003.25**

Though the first five categories of costs for this farm are the same as the other farm examples, the marketing and slaughter costs on this farm are much higher since ALL the rabbits are sold direct to the public as meat or breeding stock.

more than 100 kits per month, this business plan may start to falter unless they have an *extremely* active year-round farmer's market near a fair-size population center. Farm C may even have to visit more than one market per weekend (using spouse, kids, or hired help at alternate sites) to move all of the product. This is why we recommend a diverse marketing strategy (such as in Farm A) to decrease dependence on one market type.

So how is it that you can lose money on rabbits? Farm D is in a situation where it has no one to process its rabbits for legal meat sales, and its state doesn't allow on-farm processing. It has no contract live rabbit buyers in the region. Farm D is thus left trying to sell 100 *live* rabbits per month direct to the public. This means selling animals at farmer's or flea markets or a livestock sale barn at the least amount of profit per animal.

In the Farm D example, there is a net loss of $79.50 per month. With any increase in sales prices from what is listed (such as $12 per rabbit at the farmer's market instead of $10, and $6 at the livestock market instead of $5), there would be a slight profit. However, a more sound business plan would be to add in

Income Table Farm C

WHERE RABBITS SOLD	NUMBER OF RABBITS OR PRODUCT SOLD	PRICE PER RABBIT OR PER POUND	INCOME
Breeding stock	20	$30 per rabbit	$600.00
Farmer's market	80	2.8 pounds average, 224 pounds total sold @ $6.95 per pound	$1,556.80
Rabbit manure and worms	1 load manure	$40 per load	$40.00
5 roaster rabbits packaged for pet food	5 large rabbits sold for pet food	4.6 pounds average, 23 pounds total sold @ $4 per pound	$92.00
Rabbit livers	4 pounds	@ $4 per pound	$16.00
		TOTAL INCOME FOR JANUARY	$2,304.80

While costs are highest for this example farm, so is income, as all prices are set by the farmer and are direct to the public without a "middleman" involved. Secondary products are also easily sold due to wide customer contacts.

Cost Table Farm D (100 kits per month)

EXPENSES	DESCRIPTION	COST
Feed	pellets, hay, supplements	$500.00
Bedding	straw, nest inserts	$35.00
Utilities	rabbit barn electric and water	$60.00
Veterinary	ear mite medication	$7.00
Equipment	2 feeders replaced, marking spray, wire brush, bleach	$35.00
Marketing	printed brochures, farmer's market booth, sale barn fee	$95.00
Slaughter/ processing	none	$0.00
Gas costs	300 miles with 20 miles per gallon = 15 gallons @ $2.50 per gallon	$37.50
	TOTAL COSTS FOR JANUARY	**$769.50**

This farm has the same five "fixed" costs as the other examples (feed, bedding, utilities, veterinary, and equipment), but has no slaughter costs as they do not have a nearby processor to use. Since they also lack a contract rabbit buyer, they incur marketing and gas costs as they must sell all their rabbits live (direct to the public).

Income Table Farm D

WHERE RABBITS SOLD	NUMBER OF RABBITS OR PRODUCT SOLD	PRICE PER RABBIT OR PER POUND	INCOME
Farmer's market (live animals)	30	$10 per animal	$300.00
Livestock market (live fryer animals)	70	$5 average per animal	$350.00
Rabbit manure	1 load manure	$40 per load	$40.00
		TOTAL INCOME FOR JANUARY	**$690.00**

No legal processor for rabbit meat and no contract live rabbit buyers means this farm must try to sell 100 live animals per month at the most unpredictable and least profitable venues. This rabbit farm does NOT make a profit.

the nominal marketing costs of a website and sell a percentage of the rabbits as breeding stock to increase the profit margin, as in example Farm E.

Farm E has a profit of $410.50 per month ($1,190 - $779.50 = 410.50), not much for an operation of this size, but at least the enterprise is in the black now compared to Farm D—just by adding a website and some breeding stock sales. There may be some problems with this business plan, as local rabbit growers realize they can buy live rabbits at the farmer's market or livestock market and use them for breeding stock rather than pay Farm E higher prices for selected breeders. With strong Internet marketing, however, this farm should still be able to sell enough breeders to keep it profitable.

We have shown with these example farms that completely different profit margins can be realized simply by varying marketing strategies, as summarized in the table on the next page:

Cost Table Farm E (100 kits per month)		
EXPENSES	**DESCRIPTION**	**COST**
Feed	pellets, hay, supplements	$500.00
Bedding	straw, nest inserts	$35.00
Utilities	rabbit barn electric and water	$60.00
Veterinary	ear mite medication	$7.00
Equipment	2 feeders replaced, marking spray, wire brush	$35.00
Marketing	website hosting, printed brochures, farmer's market booth, sale barn fee	$105.00
Slaughter/processing	none	$0.00
Gas costs	300 miles with 20 miles per gallon = 15 gallons @ $2.50 per gallon	$37.50
	TOTAL COSTS FOR JANUARY	**$779.50**

Farm E is the exact same enterprise as Farm D, but marketing costs were increased to sell breeding stock and move the farm out of the "red."

Income Table Farm E

WHERE RABBITS SOLD	NUMBER OF RABBITS OR PRODUCT SOLD	PRICE PER RABBIT OR PER POUND	INCOME
Breeding stock	20	$30 per animal	$600.00
Farmer's market (live animals)	30	$10 per animal	$300.00
Livestock market (live fryer animals)	50	$5 average per animal	$250.00
Rabbit manure	1 load manure	$40 per load	$40.00
		TOTAL INCOME FOR JANUARY	**$1,190.00**

This farm is the exact same enterprise as Farm D, *except* that by marketing and selling 20 percent of their stock as high-end breeders, this farm is now *profitable* (whereas Farm D was not).

Profit Margin Examples

EXAMPLE FARM	COSTS	INCOME	PROFIT
Farm A	865.75	2,085.20	$1,219.45
Farm B	662.00	1,495.00	$833.00
Farm C	1,003.25	2,304.80	$1,301.55
Farm D	769.50	690.00	-$79.50
Farm E	779.50	1,190.00	$410.50

By comparing five farm marketing plans, it can be seen that changing just *one* factor can lead to radically different outcomes.

SETTING PRICES

So, is the maximum profit listed in the example farms above (around $1,200 to $1,300) the best that can be expected from a 100-kit-per-month operation? Absolutely not! In fact, these examples are deliberately on the low side so that readers will understand that rabbit raising is not a "get rich quick" scheme. One method of increasing your profit is simply increasing the price you charge for your product. You will have to discover just how high you can go on prices before you begin to lose customers.

It is a delicate balance. If you are marketing meat out in the country with tons of competitive livestock—that is, your customers can easily obtain cheap beef, pork, chicken, and lamb—they may be unwilling to pay high for your rabbit. If you live near a city, however, where you can market to upscale restaurants and people at farmer's markets who are used to paying higher prices even in the grocery stores, you may be able to increase your prices and still be very competitive.

You can also increase your prices in other market outlets. For example, if you are using a contract live rabbit buyer for your fryers, you might negotiate a higher price once you have proven you are a consistent and reliable source of rabbit for him. If you have the data to back up your claims that your breeding stock is indeed selected for superior production characteristics, you may be able to get $40+ per breeder, when others only get $30.

> **The simplest way to increase profit is to raise the price you charge. This works up to a point, but eventually you will price your product too high for the market.**

In short, if you spend the time and effort to aggressively market your animals or meat, you can increase your income per animal dramatically no matter what business plan you are utilizing.

LOWERING COSTS AND AMPLIFYING PROFITS

We have demonstrated how varying the marketing strategy for your farm will change the profit margins (as in example Farms A–E earlier in this chapter), and simply charging higher prices will also increase profits (up to a point); next, we will look at ways to work on lowering costs to increase profits.

At the beginning of this chapter, we listed eight cost categories for most small or mid-size operations: feed, bedding, utilities, veterinary, equipment, marketing, slaughter/processing, and gas. (Larger commercial growers will have to add in farmhand salaries and related costs). We have shown how utilizing different markets for your product will alter the marketing, slaughter/processing, and gas cost categories. Bedding, veterinary, and small equipment expenditures will not change appreciably with a healthy 100-kit-a-month operation. That leaves utility and feed costs to look at for savings.

Utility costs are not something you can substantially change once you have decided on a housing plan but should be carefully considered when you begin your operation. Start-up costs will be discussed later in this chapter, but the utilities will be a result of that initial housing decision. If a farmer chooses to set up a fully climate-controlled barn with heat and air conditioning, utility costs will be vastly different from a farm that uses fans, shaded locations, breezeways in summer, and heated automatic watering systems in winter. These farms will also differ from one with non-insulated barns, utilizing large-space heaters to keep the ambient air temperature just high enough to keep water lines from freezing. The combination of the ambient temperatures of your geographic region and your rabbit housing decisions

Feed is the single highest cost of any rabbit ranch. To contain this expenditure, use genetic selection, terminal sires, feed supplements, and the culling of unproductive breeders.

will impact utility costs every year for as long as you are in operation. Predicted utility expenditures will require careful study before you even begin your rabbitry.

As mentioned throughout this book, feed costs are the single highest expense on a rabbit farm (often more than all the other costs combined) and the most important cost to contain. The old livestock farmer adage "you can't starve a profit from your animals" is quite true, however! You do not want to skimp on amount or quality of feed. So how do you lower this cost?

First, when selecting breeding stock, you need to look at factors such as the doe's milking ability, how many kits in a litter survive, how fast their kits grow, and how efficiently the kits turn feed to meat. Moving any one of these factors in a positive direction can create a significant reduction in feed costs. If a doe is a heavier milker, giving the kits a better start and increasing immunity, the result will be faster growth and more kits surviving. If the kits themselves are more feed-efficient after weaning (needing less feed per pound of weight gained), hit market weight a week sooner, or gain an additional half pound in the same amount of growing time, you will significantly impact your feed costs or the amount of meat you produce for that feed cost.

I encourage you to refer back to Chapter 6—Selecting for Economically Important Traits—as you work on lowering feed expenditures. Using terminal sires (see Chapter 2) will also contribute to your net profit due to their offsprings' naturally faster growth rate and hybrid vigor. Adding feed supplements, such as Calf-mana or sunflower seeds, to raise the fat percentage of the ration will contribute to the kit's feed efficiency—lowering feed costs (see Chapter 4).

Another factor in curtailing overall operation costs is culling nonproductive does. Suppose you breed 15 does in a month, and 3 do not get pregnant, 1 doesn't make a nest and the kits die, and 3 have small

Worms are a lucrative sideline venture and are handy helpers with manure disposal.

litters of 3 kits (while the others have 8). In this case, you will only produce 73 kits. After culling the problem stock, when you re-breed, 1 doesn't get pregnant, 1 doesn't make a nest, and 1 litter is small with 3 kits, but other does have more than 8 kits each and you can foster them to the dam with the small litter—bringing her litter up to 8 kits. Now the farm produces 104 kits (31 more). Of course, you will spend more on feed to raise the additional kits, but with 104 animals to sell (versus 73 kits), there is more income to cover the fixed monthly costs of gas, utilities, bedding, veterinary, small equipment, and marketing for the farm.

Do not forget those sideline products to lower your costs or increase your income. Rabbit manure is a super soil enhancer. By loading old feed bags with dry manure you can sell it through farmer's markets, garden centers, and plant nurseries, in keeping with sustainable agriculture values. A couple of bags may fetch enough to buy a bag of rabbit feed. (What goes in must come out . . .) Worms raised under the cages or in compost storage areas can be sold at the same outlets, and also as bait if you live near a popular fishing spot. Neither of these products costs you anything except a bit of sales time and maybe some deliveries.

Some rabbit-raising books advocate selling rabbit pelts as a major sideline product and even spend several chapters describing how to skin and tan properly. We have not included this because, if you are having your fryers slaughtered at a processing plant, they will not take the time to skin without damaging the hides. They work way too fast to be concerned about the skinning. If you are slaughtering them yourself, you can take care with the hides, but tanning is very time consuming, and the hides at the age where rabbits are slaughtered for meat are too thin to be considered garment quality. They can be used for crafts, though, and if you learn to tan well, they could retail for $3–$7 via Internet sales or at your farmer's market booth. Some hobby stores or novelty shops will buy wholesale hides for $2–$3 per pelt. Older breeding rabbits will have thick enough hides to make a higher-grade pelt and fetch the best price.

We never tried to sell lucky rabbit's feet, which can often be sold at the same outlets as the hides. They are not difficult to prepare, however (much easier than tanning). Some kids absolutely love them, and others are creeped out by them, but even adults will buy them for good luck charms. I have seen them retail for between $1–$2 each.

START-UP EXPENDITURES

As mentioned in Chapter 8, regulations regarding on-farm processing vary widely from state to state in the United States. Therefore, providing information on costs for setting up and running a processing facility is pretty much impossible. Unless you have no other options, we do not recommend that you try to set up your own processing facility at the same time as you are *initiating* rabbit farming and figuring out how to market your rabbits. Taking on too much at once is the surest way to do none of it well. If there is any possible way to avoid packaging rabbit yourself—at the onset—you can keep your start-up costs much lower while you learn the industry. Later, when your markets are solid, setting up to process your own and even other raisers' rabbits may move you from a part-time to a full-time farmer.

Even without considering processing, you still have to calculate the start-up costs of your rabbit ranch and how long it might take you to recover the initial capital investment and realize a *real* profit for the enterprise. The first thing you have to do is determine the number of cages you will have, so you can calculate the barn size or land area you will need. This was described in Chapter 5 in detail.

Once you have decided on the number of cages, plan your barn(s), water systems, and electricity. These will vary depending on the weather in your part of the country. Refer to Chapter 5 on the need to keep the animals cool, their environment clean and well ventilated, the water flowing, and enough lighting for easy breeding and good growth.

With all of these factors in mind, look back to the earlier text on making a profit and carefully analyze the marketing prospects in your area before you invest one dollar. Suppose you decide that you might be able to make $1,000 per month profit from a 100-kit operation, and you want to set up your farm with prebuilt cages, "top of the line" insulated barns with concrete floors, a sewer system, and ducted heat and air. You also plan to hire a company to run the electricity and watering systems. In that case, you could easily run your start-up costs to a point where it would take you a lifetime for it to pay off! You need to

The first step is to calculate the number of cages you will need for your rabbit ranch vision.

Make a stab at calculating projected profits *before* you build that dream barn!

"Do it yourself" philosophy is key to keeping down your initial capital investment.

carefully price check all of your equipment and housing plans to keep your initial outlay down to a more reasonable level, even if it means a slightly smaller start-up. You can always invest further once you see how well your market grows. One option is to build the cages yourself, run your own water lines, use simple pole barns, or convert existing barns for rabbit housing.

How much you allow for start-up costs depends on your projected markets. If you are shooting for high-end restaurants, high-volume Internet sales, or high-dollar breeding stock, you can invest more than if you have a set-price contract rabbit buyer or are relying on small local farmer's markets and sale barns, where the price you might be able to command will be lower. Though every farm will be different, you can reference Chapter 11 for a table enumerating our rabbit start-up costs at Chigger Ridge.

NOTES ON TAXES AND INSURANCE

DISCLAIMER: We are not tax experts, but we will explain enough so that you can make decisions and keep the necessary records. Most small and mid-size farms in the United States are not a separate tax entity (even if they are an LLC). Your farming activities will likely be included as part of your personal income tax obligations. You need to keep receipts for everything you buy related to farming and put a note on it at the time of purchase stating what it is for.

Eventually, you will want to organize these receipts into categories, such as the eight listed in the cost tables earlier in this chapter, and

SCHEDULE F
(Form 1040)

Department of the Treasury
Internal Revenue Service (99)

Profit or Loss From Farming

▶ Attach to Form 1040, Form 1040-SR, Form 1040-NR, Form 1041, or Form 1065.
▶ Go to *www.irs.gov/ScheduleF* for instructions and the latest information.

OMB No. 1545-0074

2021

Attachment
Sequence No. **14**

Name of proprietor

Social security number (SSN)

A Principal crop or activity

B Enter code from Part IV ▶

C Accounting method: ☐ Cash ☐ Accrual

D Employer ID number (EIN) (see instr.)

E Did you "materially participate" in the operation of this business during 2021? If "No," see instructions for limit on passive losses ☐ Yes ☐ No
F Did you make any payments in 2021 that would require you to file Form(s) 1099? See instructions . . . . ☐ Yes ☐ No
G If "Yes," did you or will you file required Form(s) 1099? . . . . ☐ Yes ☐ No

Part I Farm Income—Cash Method. Complete Parts I and II. (Accrual method. Complete Parts II and III, and Part I, line 9.)

1a	Sales of purchased livestock and other resale items (see instructions)	1a	
b	Cost or other basis of purchased livestock or other items reported on line 1a . . .	1b	
c	Subtract line 1b from line 1a	1c	
2	Sales of livestock, produce, grains, and other products you raised	2	
3a	Cooperative distributions (Form(s) 1099-PATR) .	3a	3b Taxable amount . . .	3b	
4a	Agricultural program payments (see instructions) .	4a	4b Taxable amount . . .	4b	
5a	Commodity Credit Corporation (CCC) loans reported under election	5a	
b	CCC loans forfeited	5b	5c Taxable amount . . .	5c	
6	Crop insurance proceeds and federal crop disaster payments (see instructions):		
a	Amount received in 2021	6a	6b Taxable amount . . .	6b	
c	If election to defer to 2022 is attached, check here . . . . . . ▶ ☐	6d Amount deferred from 2020	6d	
7	Custom hire (machine work) income	7	
8	Other income, including federal and state gasoline or fuel tax credit or refund (see instructions)	8	
9	**Gross income.** Add amounts in the right column (lines 1c, 2, 3b, 4b, 5a, 5c, 6b, 6d, 7, and 8). If you use the accrual method, enter the amount from Part III, line 50. See instructions . . . . ▶	9	

Part II Farm Expenses—Cash and Accrual Method. Do not include personal or living expenses. See instructions.

10	Car and truck expenses (see instructions). Also attach Form 4562	10		23	Pension and profit-sharing plans . .	23	
11	Chemicals	11		24	Rent or lease (see instructions):		
12	Conservation expenses (see instructions) .	12		a	Vehicles, machinery, equipment . .	24a	
13	Custom hire (machine work) . . .	13		b	Other (land, animals, etc.)	24b	
14	Depreciation and section 179 expense (see instructions)	14		25	Repairs and maintenance	25	
				26	Seeds and plants	26	
15	Employee benefit programs other than on line 23	15		27	Storage and warehousing	27	
				28	Supplies	28	
16	Feed	16		29	Taxes	29	
17	Fertilizers and lime	17		30	Utilities	30	
18	Freight and trucking	18		31	Veterinary, breeding, and medicine .	31	
19	Gasoline, fuel, and oil	19		32	Other expenses (specify):		
20	Insurance (other than health) . .	20		a	....	32a	
21	Interest (see instructions):			b	....	32b	
a	Mortgage (paid to banks, etc.) . .	21a		c	....	32c	
b	Other	21b		d	....	32d	
22	Labor hired (less employment credits)	22		e	....	32e	
				f	....	32f	

33	**Total expenses.** Add lines 10 through 32f. If line 32f is negative, see instructions . . . . ▶	33	
34	**Net farm profit or (loss).** Subtract line 33 from line 9	34	
	If a profit, stop here and see instructions for where to report. If a loss, complete line 36.		
35	Reserved for future use.		
36	Check the box that describes your investment in this activity and see instructions for where to report your loss:		
a	☐ All investment is at risk.	b ☐ Some investment is not at risk.	

For Paperwork Reduction Act Notice, see the separate instructions.   Cat. No. 11346H   Schedule F (Form 1040) 2021

As a farmer, you will likely be filing a Schedule F form along with your personal income 1040. Look it over to see what records you need to keep.

then tabulate how much was spent on feed, bedding, equipment, veterinary supplies, slaughter costs, and marketing for the year—with receipts to back up *everything*. You will need to record mileage for farm-related business if you drive the vehicle for personal use as well. You can also take a percentage of rabbit-related utilities and maybe some office space as deductions, but do not get greedy with this. Taking 50 percent of your water, electric, or phone bill as shared rabbit expenses will be considered excessive and trigger an IRS red flag. Be even more cautious when claiming office space.

You also need to keep receipts from all the sales you make every month, whether it is to a business such as a restaurant, an individual at the farmer's market, or a customer purchasing breeders. You might not always have a person's name to put on every receipt, like at a farmer's market, but you still need to record the date, location, and number of rabbits you sold that day and the price you received.

Start-up costs, such as barns or sheds, cages, bulk purchase of nest boxes and feeders, or running electric or water to your barn, can be separated out over several tax years (and depreciated). This is advantageous in that you don't have to take a huge loss your first year when all of these costs are incurred, but you don't have a lot of sales yet. You can take part of these start-up costs off your taxes in later years to offset your increased income as your markets grow.

You should talk to a professional tax preparer about the difference between a "hobby farm" and a "working farm." You cannot take anything off your taxes as an income loss if your rabbit raising is just a hobby and you don't approach it from a professional standpoint with the *intent* to make money. Taking losses off your taxes for an enterprise that is not truly a business can get you in trouble with the IRS.

One way to help document that you are approaching things as a professional farmer is to write up a business plan (and be willing to change it from time to time if you are not making a profit). If you don't know how to do this, your county farm extension agents should be able to help you. On the other hand, claiming to be a hobby farmer and not taking any costs off your income taxes doesn't keep you out of trouble if, in reality, you are making a profit over those costs.

Receipts of income are just as necessary to maintain as your expenditure logs. They can be simple or fancy, as long as all sales are dated and the amount received recorded faithfully.

You must claim earnings on your income taxes from *any* source. Even if you did not make any actual profit during your initial year or two due to start-up expenditures, you should begin claiming your deductions on day one if you plan to call yourself a farmer later.

Luckily, many states do not require farmers to collect sales tax on food being sold direct to the public, but check to be sure! If you are selling nonfood items, however, you may need to collect sales tax on these and turn them over to the state on a regular basis. It is better to go through the red tape to find out what you legally need to do rather than get into trouble with some bureaucrat with the power to fine you.

Finally, you should consider what you are comfortable with as insurance. If you are selling all of your rabbits to a live rabbit buyer or at a sale barn, you might not need to have any insurance at all. If you are selling a lot of breeding stock and have on-farm visitors, contact your insurance agency to see if your home insurance will cover any injuries. You do not want them to decide after an accident that your liability won't cover it since it was for business.

If you are selling rabbit meat to people, product liability insurance is advisable. Some farmer's markets actually require a million-dollar product liability policy. This sounds like a lot, but don't panic; we found that some farm-friendly insurance agencies will write an umbrella policy (meaning it follows you everywhere you go and covers you like an umbrella) for a million dollars that only costs $100–$200 per year. Just make sure they specifically write in product liability. This type of insurance will cover all of your needs, from someone breaking a leg on your farm by tripping

> **Talk to a professional tax preparer about the difference between a "hobby farm" and a "working farm."**

over a rock, to a person slipping in your farmer's market booth, to someone claiming to get sick on your rabbit.

You are not covered if you are negligent, however! Stretching an electric cord across a busy walkway at the farmer's market, leaving tools around for people to trip over at your farm, not having a "safe handling instruction" label on your rabbit, or not having a thermometer in your freezers to make sure they are at the proper temperature can all invalidate your insurance!

Do not get discouraged! People who like to raise animals and work outside may find all the suggested paperwork, documentation, analysis, and regulations a bit overwhelming. Unfortunately, successful farming requires that you wear many hats. You must be able to take care of your animals but also be a maintenance worker, public relations specialist, marketing expert, and accountant. Successful farmers are not dumb or lazy! But I promise, once you have done it for the first year, it becomes much easier, and you can concentrate on the rabbits. Remember, if you raise your rabbits in a healthy environment and follow the advice in this book, you can provide a great food source for your friends, family, and community. You *can* do this while making a profit. The rabbit is completely geared toward doing its part if you do yours!

The rabbit's very nature is geared toward providing a predictable and dependable farm income.

CHAPTER 11

ALTERNATIVE AGRICULTURE IN ACTION

CHIGGER RIDGE RANCH: A RABBIT FARM EVOLUTION

Even though every farm is unique, many folks want to know "how we did it." How did we keep costs down and increase profits? How did we change our business plan to grow in this unusual agriculture endeavor? How much time did we have to invest to be successful? To answer these questions, we briefly describe the evolution of our enterprise.

We started with one small barn of 20 cages to learn the intricacies of raising rabbits. It was quickly evident that this would become a successful part of our farm, with a more predictable income than from our lambs, which could be affected by droughts, predators, and parasites. Therefore, we decided to take the plunge and buy our start-up items in bulk to save money by utilizing discount offers on large purchases and reduced shipping costs. This also allowed us to apply for (and receive) alternative agriculture grant money.

There are many such programs, both federal and state, to help farmers build a profitable business (contact your county Agriculture Extension Service Offices for help finding them). On page 180, you'll see a table with our start-up costs for 100 cages (adjusted to 2019 prices). This will vary depending on choices of wire type, water systems, number of cage bank rows, etc., but this will give you some idea of what to expect and what items you need to price out for your farm.

We built one barn ourselves; the others were converted for rabbits from old barns already on the farm. We built all of our own cages and rented trenchers to run the water

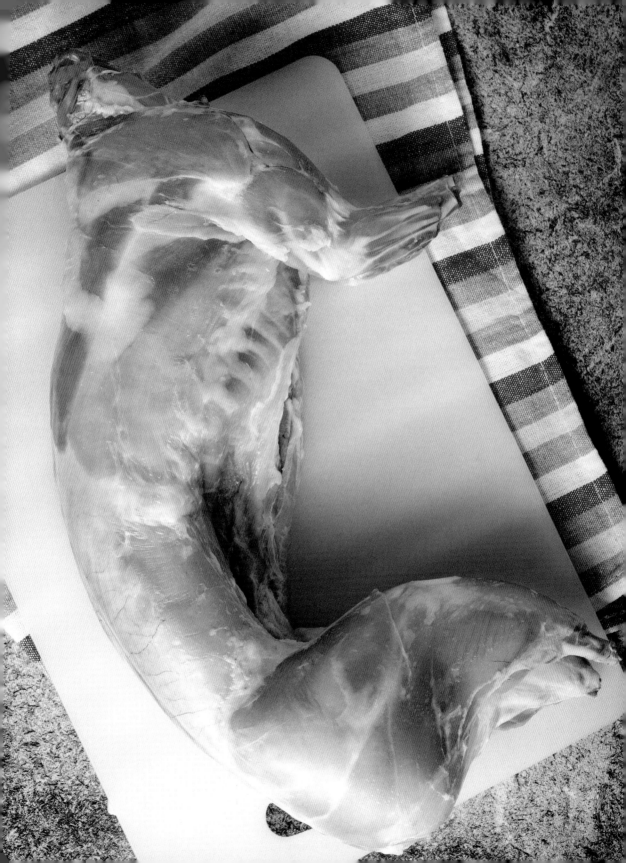

Start-up Costs for 100 Cages

ITEM DESCRIPTION	NUMBER TO BUY	COST PER UNIT	TOTAL
Nest boxes Universal Sani-Nest	50	12.7	$635
Feeders 9.5" Fine-X	100	5.45	$545
Hay racks 11.5" (each shared between 2 cages)	50	5.25	$262.5
Feeder covers	10	1.9	$19
Resting boards	100	2.25	$225
Water valves	100	3.5	$350
Pipe saddles for water valves	100	1	$100
Thermo cubes for water systems	6	16.95	$84.75
Water system	6	225	$1,125
1" x 2" 18" side wire, 16g	1000 feet	0.88/ft	$880
1" x 2" x 24" roof wire, 16g	150 feet	1.17/ft	$175.5
1" x ½" x 24" floor wire, 14g	150 feet	1.47/ft	$220.5
Door hangers	100	0.45	$45
Door latches	100	1.35	$135
cage clips	box 10,000	62.8	$62.8
PVC ½" plus adaptors, elbows…	300 feet plus		$100
Freezers	4	170	$680
Digital scales	2	110	$220
Sub Total			**$5,865.05**
Bulk order discounts 10%			-$586.5
Shipping			$500
Sub Total			**$5,778.55**
State Alternative Agriculture grant (30%)			-$1,733.55
Total Costs			**$4,045**

and electric lines. It is vital to keep start-up costs down while your markets grow. If you do not have any barns or sheds on your property, remember that even simple pole barns will work, and walls can be built of wire covered with burlap and plastic.

We began our marketing with a USDA-approved meat processor who would package our rabbits on weekends for sale at our local farmer's market. This was possible, as our state did not require carcass-by-carcass inspection. Later we moved to a USDA-inspected Mennonite processor who catered to nontraditional and smaller farms. We were already selling lamb direct to the public, so we only had to purchase a few more freezers for rabbits. (We transported the meat frozen in freezers hooked to an inverter from the truck battery so that we did not have to purchase refrigerated trucks.)

Knowing we wanted to expand, and worried that the farmer's market might not have enough customers for a larger operation, we put in for a "grower number" with a major regional rabbit contract buyer. They would pick up our live animals at a nearby location on a set day and transport them to the processing plant, where they were packaged and distributed nationally. We eventually got our number, and they bought all we wanted to raise.

We kept up with our farmer's market sales, however, to maintain our diversity and to sell some product at a higher price than the contracted buyer would give us. It was a good thing, too, for when gas prices skyrocketed, our rabbit buyer began cultivating growers nearer to his location and cancelled the more distant farms. By this time, however, we had been approached by chefs and began distributing to restaurants and food markets. Also, our rabbits were now producing well enough for us to sell breeding stock with confidence. We increased our

An old Tennessee tobacco barn in a nice shady spot, "repurposed" for rabbits.

"Sweat equity" pays off . . . eventually.

We varied our markets early on to increase profit and dependability.

income even more by selling cull breeders, runts, and dead pinkie rabbits as pet food. NO rabbit, unless it was sick, was wasted!

We never really had the time to fully explore the markets for secondary products (like worms for fishing), but we did sell manure by the truckload or sometimes bartered manure for farm labor (having the buyers clean out our barns). Saving time became more important than added income as our operation grew. In retrospect, these sideline products may have increased our profits substantially, and we should have put more time into moving them.

As our markets expanded (rapidly), we eventually limited our operation to around 450 individual rabbits at any one time. Our goal was to keep 150 working does, bucks, and brood stock for sale, 150 unweaned kits (per month), and 150 "grow-outs" (per month). This limitation on rabbit numbers was strictly due to time constraints without hired farm help—our markets could have supported far more. (We constantly had to turn down business.)

TIME INVESTMENT

We were both working full time when we began with rabbits, and all of our cages were designed with time saving in mind. Automatic water is an absolute must for an operation of 450 animals (cleaning and filling that many water bottles would have taken half the day). We also found that the rabbits raised over worm beds required far less time than cages that had to have manure washed out from underneath (although these wash-out systems were better for farm tours, as they were "prettier"). Hay and pellet feeders located so cage doors didn't have to be opened and closed and walkways between cages that allowed for a wheeled cart helped to speed the daily feeding. And, still, daily feeding chores took about an hour for one person. This was our only "must do every day" chore. Going away from the farm for longer than overnight meant hiring someone for feeding.

If only it were just feeding! Rabbit raising is time-intensive. Below is a list of the time usually spent per week on husbandry and marketing activities at Chigger Ridge for our 150 fryer-per-month operation:

- 7 hours/week – Feeding
- 7 hours/week - Cage cleaning (done on a rotating basis)
- 5 hours/week - Breeding, weaning kits, putting in or checking nest boxes

- 9 hours/week - Weighing and sorting fryers and maintaining records
- 4 hours/week - Marketing phone calls, emails, farm tours, website updates
- 3 hours/week - Manure handling, worms, compost care
- 3 hours/week - Preventative medical checks on animals (teeth, feet, ears, etc.)

This adds up to **38 hours per week** spent on necessary activities. Then you have to add in the time for the less regular tasks. (Oh yes, there is more . . .) It is good to occasionally move out all animals from a barn and disinfect it (especially

Analyze the business and adjust the plan.

in between fryer groups). Periodically, doe barns need to have animals moved out of cages and the hair (that gets on everything from making nests) burned or scrubbed off. Then there are the trips to the feed store, processing plant, restaurants, supermarkets, and farmer's markets. Two people can do it even while working if they have somewhat flexible schedules, are willing to sleep less, and are committed. Or one person if they worked at it full time. (But I sure wouldn't have wanted to take on more rabbits without help!)

COST ESTIMATES

At the beginning of our rabbit farming operation (when we were raising 100 kits per month), it cost about $7.42 to raise a fryer, including *all* of the costs that each fryer had to bear: food (and the food of all the breeding stock), bedding, its share of utilities, veterinary costs for any rabbit, basic farm supplies, marketing costs, and gas (all the costs listed in Chapter 10).

Add up all these monthly costs and divide by the number of kits you have for sale that month, and you have the monthly cost of raising an individual rabbit for sale. This is the dollar amount that must be surpassed for each rabbit sold to realize a profit. When selling meat direct to the public, you also have to add in around $2.75 per rabbit as a cost for processing them at a USDA facility (but obviously this cost is not incurred when selling to a contract buyer).

As our farm evolved, we improved husbandry and genetics to increase kit numbers and survival. By raising 150 kits per month instead of 100, each kit's share of the cost of non-food items drops. In our case, this change went from $2.42 to $1.61. Therefore, just this move to a slightly larger operation reduced our cost to raise a kit from $7.42 to $6.61 each. We then

Cost per Kit Raised (at the start)	
Feed	$5.00
Bedding	$0.35
Utilities	$0.60
Veterinary	$0.07
Small Equipment	$0.35
Marketing	$0.55
Gas	$0.50
Total	**$7.42**

moved this $6.61 cost closer to and below $6.00 by selecting for does that milked heavily and kits that were more efficient at utilizing feed. We used Altex and Flemish Giant terminal sires to hit market weight faster. We also were able to buy feed in bulk as our operation grew (for a *considerable* cost savings). We lowered utility costs with more efficient water line heaters and the use of thermocubes to turn on the heaters only when needed.

> As our farm evolved, we improved husbandry and genetics to increase kit numbers and survival. Then we utilized terminal sires and selection to further reduce feed expenditures. This allowed us to decrease our cost of raising a kit from $7.42 to below $6.

Our rabbit sale price varied with our target market. Breeding stock sales were by far the most lucrative, but to maintain our reputation, we only sold the top percentage of productive rabbits as breeders. We sold both Altex and New Zealand Whites and had a waiting list of months (suggesting we could have charged more than we did). Below are the typical prices that we received per rabbit, from our highest to lowest market. These prices will not necessarily predict what you can charge for your rabbit. It depends on your location, customer base, and current economic conditions, such as inflation. We obviously began concentrating on high-end restaurants for meat sales. Note that we did not utilize livestock sale barns at all, where we may have broken even or even lost money depending on the buyers present.

- $60 per Altex buck breeder
- $40 per Altex doe breeder
- $30 per NZW doe or buck breeder
- $18.20 restaurant sale (preferred 2.8 lbs carcass weight @ $6.50/lb)
- $16.25 farmer's market fryer (preferred 2.5 lbs carcass weight @ $6.50/lb)
- $13.75 food store at wholesale (preferred 2.5 lbs carcass weight @ $5.50/lb)
- $13 farmer's market introductory weight (2.0 lbs carcass weight @ $6.50/lb)
- $11.50 rabbit contract buyer at highest weight (5.75 lbs live weight @ $2.00/lb)
- $9.50 rabbit contract buyer at lowest weight (4.75 lbs live weight @ $2.00/lb)

We set prices low enough that we could easily move all of our product but never lose money on a fryer. As we reduced costs and moved to the most profitable sale venues, our average was $8.50 profit per meat fryer each month, with a substantial increase in monthly income generated by our breeding stock sales. We also developed a nice market for *all* of our cull breeders as pet food, and since they are so heavy when slaughtered, this often resulted in several hundred additional dollars per month.

PRACTICING SUSTAINABLE AGRICULTURE

The term "sustainable agriculture" is thrown around a lot. We did not define it earlier because everyone seems to have a different idea of what it really means. This is one instance where we actually agree with the US government's definition of something. They say it quite well, in fact, in US Code Title 7 on agriculture (chapter 64, subchapter 1, section 3103). See page 185 for this description.

With its prolific breeding capability, excellent feed conversion, and dependable economic potential, the rabbit fits this definition very well. At Chigger Ridge we embraced the "integrated system" approach even further by combining bees, rabbits, sheep, and planted field crops of sunflowers and buckwheat to the benefit of all.

We would also like to tip the hat to the "slow food" movement, which states that food should be grown locally, prepared with care, and consumed with appreciation!

SUMMARY

Raising rabbits successfully as a business or even to supplement your family's diet in a cost-effective manner means following the cardinal rule of any livestock enterprise: **You can't manage what you don't measure.** Recording all aspects of your operation is the key to profitability. Weighing your rabbits on a regular basis will tell you if your feed is quality, if your kit growth rate is adequate, and

Keep a variety of market options open with quality product suited to each customer's needs, but concentrate your efforts on your most dependable customers who provide the best rate of return.

if your does are milking well and recovering from lactation in a timely manner. Sudden weight loss may indicate a disease in time to prevent any spread.

Recording your litter sizes and dates and the kit growth rates will tell you which does might need to be culled and how productive your overall genetics are. Keeping careful track of parentage prevents too much inbreeding and tells you when you need to obtain new bloodlines. Recording costs and income tells you if you need to alter your marketing or change the price of your product. But you have to do more than simply log all this information; you must also *analyze* and *act* on the data. We tried to convey not only how, when, and what statistics you need to collect, but the changes to be made to improve these numbers and enhance the profitability and the efficiency of your enterprise.

Sustainable Agriculture is an integrated system of plant and animal production practices having site-specific application that will over the long term:

A. satisfy human food and fiber needs;

B. enhance environmental quality and the natural resource base upon which the agriculture community depends;

C. make the most efficient use of nonrenewable resources and on-farm resources and integrate, where appropriate, natural biological cycles and controls;

D. sustain the economic viability of farm operations; and

E. enhance the quality of life for farmers and society as a whole.

Sustainable Agriculture

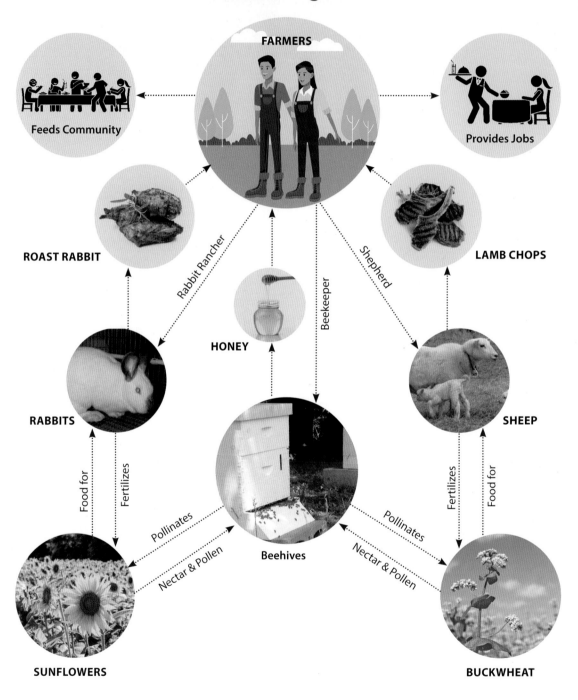

FARMERS

Feeds Community

Provides Jobs

ROAST RABBIT

Rabbit Rancher

HONEY

Beekeeper

Shepherd

LAMB CHOPS

RABBITS

Food for

Fertilizes

Beehives

Fertilizes

Food for

SHEEP

Pollinates

Nectar & Pollen

Pollinates

Nectar & Pollen

SUNFLOWERS

BUCKWHEAT

At our farm, we promote the concept of sustainable agriculture through interdependence.

With the rabbit's natural prolific nature, high feed efficiency, and valuable by-products, it is an ideal animal for the practice of sustainable agriculture. Its low fat and cholesterol and high protein content, derived mainly from high omega-3 sources of feed, make it one of the healthiest meats on the planet.

For those of you who want to raise and market rabbits successfully, we point out the following:

- Rabbits all by themselves have everything it takes to produce pounds and pounds of healthy meat. They are naturally very efficient, prolific, and therefore cost-effective.

- Because of its physical and social needs, a domestic rabbit must live in its own cage or hutch to stay happy and healthy. This means the rabbit keeper has to provide proper food, fresh water, nesting material, and toys as needed, and keep them clean, safe, and comfortable at all times. Practicing humane farming can be a time-intensive operation.

- Rabbits in close proximity can spread diseases to each other and are difficult to treat successfully. It is absolutely essential that new rabbits are quarantined and all rabbits are carefully observed on a daily basis. Fortunately, most of the "diseases" in rabbitries are actually management-related. Improper feeding and watering, incorrectly designed or placed housing, lack of ventilation, poor pest control, or unsanitary conditions contribute to most of the health issues in rabbitries.

- Measure, Analyze, Act! With their very rapid turnover rate, small changes in herd genetics and management can lead to easily seen positive or negative results in a short amount of time. A rabbitry without good documentation of breedings, litter sizes, kit survival rates, and rate of growth, and which does not cull and replace animals with improved stock regularly, is going nowhere fast.

- Despite all the advantages of the rabbit for providing affordable and extremely healthy meat, the industry is in its infancy in the United States. In many cases, it is up to the farmer to create markets and wade through often hazy regulations. The lack of widely available, federally inspected meat rabbit processors is a major bottleneck for the industry, and the rabbit rancher may have to work at locating legal alternative processors.

- The successful rabbit rancher must be prepared to analyze his or her income versus expenses on a regular basis to be able to set a price that will produce a profit, or make changes that will do so. This book is unique in that it provides the reader with information on the economics as well as the husbandry of rabbits.

When raised in accordance with the precepts of this book, rabbit can be a low-cost source of nutritious food for a family or a means to additional income on almost any scale. Indeed, when attention is paid to animal health and sanitation, and adequate time is invested in the measurements described above, it is difficult to *not* make a profit with rabbits!

REFERENCES
AND ADDITIONAL INFORMATION
(Electronic sources were all accessed March 23, 2018.)

Animal Welfare Act and Animal Welfare Regulations from USDA, APHIS 41-35-076. Part 3, Subpart C, Section 3.50–3.66, pages 153–164. 2017, January 1. Retrieved from: https://www.aphis.usda.gov/aphis/home/. Accessed at: https://www.aphis.usda.gov/animal_welfare/downloads/AC_BlueBook_AWA_508_comp_version.pdf.

ARBA (American Rabbit Breeders Association) Recommendations for the Care of Rabbits and Cavies. Retrieved from: arba.net. Accessed at: https://www.arbadistricts.net/PDFs/CAW.pdf.

Arrington, L., Platt, J., and Franke, D. (1974). "Fat Utilization by Rabbits." *Journal of Animal Science* 38(1):76–80. February 1974. Accessed at: https://www.researchgate.net/publication/18312520_Fat_Utilization_by_Rabbits.

Grannis, J. (2002). U.S. Rabbit Industry Profile. USDA:APHIS:VS Report. June 2002. Retrieved from: aphis.usda.gov. Accessed at: https://www.aphis.usda.gov/animal_health/emergingissues/downloads/RabbitReport1.pdf.

Hylton, H. (2012). "How Rabbits Can Save the World (It Ain't Pretty)." *TimeWorld.* 2012, Dec. 14. Retrieved from: http://time.com/vault/. Accessed at: http://world.time.com/2012/12/14/how-rabbits-can-save-the-world-it-aint-pretty/.

Karr-Lilienthal, L. (2011). The Digestive System of the Rabbit. Retrieved from: http://articles.extension.org. Accessed at: http://articles.extension.org/pages/61402/the-digestive-system-of-the-rabbit.

Karr-Lilienthal, L., and Young, A. (2013). Dental Problems in Rabbits. Retrieved from: articles.extension.org. Accessed at: http://articles.extension.org/pages/33007/dental-problems-in-rabbits.

Kelleher, S. (2010). "Gastrointestinal Problems in Rabbits." *LafeberVet.* 2010, July 28. Accessed at: https://lafeber.com/vet/gastrointestinal-problems-in-rabbits/.

Lane, T. J. (1999). Rabbit Production in Florida. University of Florida Cooperative Extension Service. Fact Sheet VM-51. April, 1999. Retrieved from: mysrf.org. Accessed at: http://mysrf.org/pdf/pdf_rabbit/r14.pdf.

Leary, S., Underwood, W., Anthony, R., Corey, D., Grandin, T., Gwaltney-Brant, S., Meyer, R., Regenstein, J., Shearer, J., Smith, S. A., Golab, G. C., and Johnson, C. (2016). AVMA Guidelines for the Humane Slaughter of Animals: 2016 Edition. Handling procedures for rabbits, pages: 48–50. Retrieved from: avma.org. Accessed at: https://www.avma.org/KB/Resources/Reference/AnimalWelfare/Documents/Humane-Slaughter-Guidelines.pdf.

Lebas, F., Coudert, P., de Rochambeau, H., and Thébault, R. G. (1997). Chapter 3: Reproduction. Found in: The Rabbit: Husbandry, Health and Production. Food and Agriculture Organization of the United Nations Corporate Document Repository # ISSN 1010-9021. Retrieved from: http://www.fao.org/publications/card/en/c/a465f8f1-79dc-585b-8ca5-fa06486e4d71.

Lombolt, N., Dept. of Pharmacology University of Copenhagen. The Use of Carbon Dioxide Anesthesia before Slaughter. Accessed at: http://www.butina.eu/fileadmin/user_upload/images/articles/cabon_dioxide.pdf.

Lukefahr, S. D., Cheeke, P. R., McNitt, J. I., and Patton, N. (2004). "Limitations on Intensive Meat Rabbit Production in North America: A Review." *Canadian Journal of Animal Science.* 84:349–360. Retrieved from: www.nrcresearchpress.com. Accessed at: http://www.nrcresearchpress.com/doi/pdf/10.4141/A04-002.

Lukefahr, S. D., Garza, M. T., Schuster, G. L., and McCuistion, K. C. (2012). "Meat Rabbits Finished on Sweet Potato Forage Looks Good in Texas Research." The Stockman Grassfarmer 12(5):7–9. Accessed at: https://pdfs.semanticscholar.org/347e/736d6fe9246fae259a992f72c100b75c99f4.pdf.

Manna Pro feed. Calf Manna® Ideal Performance Supplement for Rabbits. Accessed at: https://www.mannapro.com/products/rabbit/rabbits.

Mayer, J. (2016). Bacterial and Mycotic Diseases of Rabbits: Pasteurellosis. The Merck Veterinary Manual. Merck Sharp & Dohme Corp. Retrieved from: https://www.merckvetmanual.com/. Accessed at: https://www.merckvetmanual.com/exotic-and-laboratory-animals/rabbits/bacterial-and-mycotic-diseases-of-rabbits.

McNitt J., Lukefahr, S., Patton, N., and Cheeke .P. (2013). *Rabbit Production, 9th edition.* CABI Publishing.

Medellin, M. F., and Lukefahr, S. D. (2001). "Breed and Heterotic Effects on Postweaning Traits in Altex and New Zealand White Straightbred and Crossbred Rabbits." Journal of Animal Science 79(5): 1173–1178. May 2001. Accessed at: https://doi.org/10.2527/2001.7951173x.

Post, R., Budak, C., Canavan, J., Duncan-Harrington, T., Jones, B., Jones, S., Murphy-Jenkins, R., Myrick, T., Wheeler, M., White, P., Yoder, L., and Kegley, M. (Editors) (2007). A Guide to Federal Food Labeling Requirements for Meat, Poultry, and Egg Products. Food Safety and Inspection Service, USDA. Retrieved from: https://www.fsis.usda.gov/wps/portal/fsis/home. Accessed at: https://www.fsis.usda.gov/wps/wcm/connect/f4af7c74-2b9f-4484-bb16-fd8f9820012d/Labeling_Requirements_Guide.pdf?MOD=AJPERES.

Rabbit from Farm to Table. 2015, March 12. USDA Food Safety and Inspection Service Food Safety Information. Retrieved from: https://www.fsis.usda.gov/wps/portal/fsis/home. Accessed at: https://www.fsis.usda.gov/wps/wcm/connect/bcb4cfe5-4af2-4406-8ab9-4a1c8273dff5/Rabbit_from_Farm_to_Table.pdf?MOD=AJPERES.

van Praag, E. Protozoal Enteritis: Coccidiosis. Retrieved from: MediRabbit.com. Accessed at: http://www.medirabbit.com/EN/GI_diseases/Protozoal_diseases/Cocc_en.htm.

van Praag, E. Bacterial Enteritis and Diarrhea in Weaned and Adult Rabbits. Retrieved from: MediRabbit.com. Accessed at: http://www.medirabbit.com/EN/GI_diseases/Generalities/Enteritis_en.htm.

Xiong, G.-Y., Xu, X.-L., Zhu, X.-B., Zhou, G.-H., and Shi, S. (2008). "Effects of Withholding Food and/or Water Supply on the Quality of Meat from Rex Rabbits." *Journal of Muscle Food.* Accessed at https://onlinelibrary.wiley.com. Retrieved at: https://onlinelibrary.wiley.com/doi/abs/10.1111/j.1745-4573.2008.00123.x.

APPENDIX A:
RABBIT PURCHASE AGREEMENT EXAMPLE

All text in parentheses ( ) is filled in by you—the XYZ Rabbit Ranch.

(XYZ Rabbit Ranch) has received today, (DATE), the sum of (X dollars) from (Name of Buyer) as a deposit for the purchase of the following rabbits:

(X number of) NZW doe rabbits (3–6 months of age) at ($X) per rabbit
(X number of) NZW buck rabbits (3–6 months of age) at ($X) per rabbit
(X number of) Altex doe rabbits (3–6 months of age) at ($X) per rabbit
(X number of) Altex buck rabbits (3–6 monts of age) at ($X) per rabbit
(X) total number of rabbits, for ($X) total cost.

The purchaser by signing below understands and agrees to the following:

1. The rabbits are purebred but not registered.

2. Any veterinary checks or health certificates required by the purchaser will be the responsibility of the purchaser in full.

3. The purchaser must pick up the last of the rabbits at (XYZ Rabbit Ranch) at a mutually agreed-upon time between (DATE) and (DATE). After (DATE), there will be a ($X) per rabbit per day boarding fee until (DATE). After that time the deposit will be forfeited by the purchaser, as failure to pick up the rabbits will significantly impact the operation of (XYZ Rabbit Ranch).

4. The purchaser understands that the seller will do his or her best to provide the agreed-upon number of rabbits listed above, but as some are not yet born, it is impossible to guarantee the exact numbers.

5. If the purchaser decides at a later date that he wants a lesser number of rabbits than requested above, he will forfeit his deposit in full; it will not be applied toward the purchase of a lesser number of rabbits.

Purchaser: _____ Date: _____

Seller: _____ Date: _____
Seller Doing Business As (XYZ Rabbit Ranch)

APPENDIX B:
BROCHURE EXAMPLE

Below is an example of a simple trifold brochure from our farm, "Chigger Ridge Rabbits." Most word processing software programs have some sort of brochure template you can use. Feel free to copy and modify what we did to fit your farm! Below are photos of the finished product.

How much does rabbit cost?

Most of our rabbit is sold as "Fryer Rabbit" for $6.95 per pound and they are 2–3 pounds. They are young, tender, and already cut up.

Occasionally we have "Roaster Rabbits" for $5.35 per pound that are 3 pounds and up. Some folks like these larger whole rabbits for certain recipes.

Rabbit Livers are $4.00 per pound.

Where can I purchase Chigger Ridge Rabbit?

This year we are pursuing new farmer owned Farmer's Markets that are part of the "local foods" movement. Thus we plan to sell at XXX or XYZ. Check our website, call, or email for dates and times.

How are the rabbits raised?

Our rabbits are housed up off of the ground in self-cleaning hutches and provided with an automatic watering system. We feed them commercial rabbit pellets (without any antibiotics, hormones, or animal by-products). We also feed hay to provide them with a more natural food and to give them occupation. They are housed in social groups and provided with toys and sticks to chew if they want. Since rabbits do not like the heat, our rabbits are kept in well-shaded barns with lots of ventilation and are even provided with fans in hot weather.

Chigger Ridge Rabbit
Cumberland Furnace, TN. 37051
info@chiggerridge.net
XXX-XXX-XXXX

Kick the fat habit, switch to rabbit!

Chigger Ridge Domestic Rabbit

(The other other white meat).

Tel:XXX-XXX-XXXX
Web: www.chiggerridge.net
Email: info@chiggerridge.net

Why Rabbit?

The movement in the United States and elsewhere around the world is toward seeking lower fat, lower cholesterol, lower calorie foods which are still a good source of protein and yet actually taste good. Sounds impossible, right? Not if you consider one of the oldest sources of meat on the planet—the rabbit! Rabbit is a fine-grained, mild flavored all-white meat that is lower in fat, cholesterol, and calories and higher in protein than any other meat.

As world populations rise and there is less land to raise quality food, the rabbit may play an increasingly important role in our national and international food supplies. One rabbit doe can produce 250+ pounds of meat in a year and is many times more efficient than a cow in turning feed into meat.

Rabbit Meat – the heart-healthy alternative.

Rabbit meat, has the lowest cholesterol of any meat including chicken. United States Department of Agriculture (USDA) circular #549 further states that rabbit has the highest percentage of protein and the lowest percentage of fat and calories when compared to veal, chicken, turkey, lamb, beef, and pork. The percentages from this circular are presented in the table below.

Species	Calories %	Protein%	Fat
Rabbit	795	20.8	10.2
Chicken	810	20.0	11.0
Veal	840	19.1	12.0
Turkey	1,190	20.1	20.0
Lamb	1,420	15.7	27.7
Beef	1,440	16.3	28.0
Pork	2,050	11.9	45.0

Note: these numbers are for your typical U.S. feedlot finished animal. Free range or grass-finished livestock will have lower fat than that listed in the table above.

Cooking Rabbit

Rabbit meat can be prepared in any of the ways that chicken can—fried, grilled, roasted, braised, stir fried, smoked, or stewed. It is a mild flavored, fine-grained dense all white meat. One rabbit will usually feed 4–5 people. Even though they are extremely low fat, we have found they stay moist even on an outdoor grill due to the denseness of the meat. Our favorite way to cook rabbit is to simply BBQ. Unlike the cottontail, the domestic meat rabbit does not need to be tenderized by parboiling or long marinating and does not have a "gamey" taste.

Rabbit livers are like larger, sweeter chicken livers (without the antibiotics and hormones). They are usually fried but we also like them grilled. Gourmet chefs prize rabbit livers for making liver paté

APPENDIX C:
RESTAURANT FLYER EXAMPLE

Chigger Ridge Rabbit

Chigger Ridge Ranch is pleased to announce that we are now offering succulent domestic rabbit wholesale. Rabbit is a fine-grained, mild flavored, all-white meat which has been a staple around the world for generations and takes up a wide variety of spices and marinades well. Marinading for tenderness is not necessary with our product, however, as our rabbit is packaged between two and four months of age and thus guaranteed to be top-quality. We sell our rabbit whole frozen, and it is vacuum packaged at a USDA-inspected facility. When cut up, it provides two front legs similar in size to large chicken wings, two substantial back legs larger than a big roaster chicken, and two loins/tenderloins. A rabbit cut up usually feeds four people. If cooked whole and deboned after cooking, it will feed five or more. It can be cooked in virtually any way that chicken can: fried, grilled, roasted, stewed, or sautéed.

Our packaged rabbits are between 2.5 and 3 pounds, and our wholesale price is $6.50 per pound. We are able to deliver orders of 15 rabbits to your door free of charge. If you wish to order a lesser number, we can deliver for a fee but would prefer to deliver free if you can wait for your product until we are making another delivery in your area. This policy is in keeping with our commitment to conservation and sustainability.

Whether you are an establishment that caters to the exotic, continental cuisine, Asian tastes, traditional Hispanic dishes, farm-to-table meals, or just plain ol' BBQ, the rabbit can fit your menu. As a meat that is the lowest of all meats in cholesterol and fat and the highest in protein, it can also fit your heart-healthy menu options.

At Chigger Ridge we are dedicated to humane animal husbandry and sustainable local agriculture. We have strict sanitation protocols and provide for both the physical and social needs of our animals. We do not use any antibiotics or feed that contains added hormones or animal by-products.

We only provide rabbit that we raise ourselves, so we will have a limited number of restaurants that we can consistently serve. If you wish to be one of them, please visit our website at XXX, email us at XXX, or call XXX.

INDEX

S

sale barns, selling to, 143

sales. *See* commercial rabbitry; markets and marketing

sanitation and disease prevention, 84

"scruffing" rabbits, 27

selection and genetics, 96–111

  conversion ratio (overall) and, 104–5

  culling and, 111

  doe milk production and, 104

  glaucoma gene/selection illustration, 99–102

  inbreeding, line breeding, outcrossing, and crossbreeding, 96–99

  kit feed efficiency and, 105–6

  kit growth rate and, 106, 107

  marking and identification, 108–10

  percent of genes shared between relatives, 98

  practical selection strategy, 106–8

  Punnett Square and, 100–101

  records, 110

  selecting economically important traits, 102–8

selling meat. *See* commercial rabbitry; markets and marketing

sexing rabbits, 28–29

skin diseases, 38–39

slaughter and carcass

  approaching processors and, 139

  carcass quality, 118

  comparison of poor and high-quality carcasses, 119, 125

  cooking rabbit and, 126–27

  cost accounting. See finances

  cuts of rabbit and photos, 118–26

  methods of slaughter, 114–16

  transporting to processor, 116–17

snakes, protection from, 83

snuffles, 33–34

social needs of rabbits, 88

soiled paws, 93

sore hock, 38, 39

stores, selling to. *See* markets and marketing

sunflower seeds, black oil, 70, 71

supplements to consider, 70–71

sustainable agriculture, 10–11, 184–87

T

tattoos, 108, 109–10

taxes and deductions, 174–77

teeth, problems with, 36–38

terminal sires, 23–26, 75, 171, 184

time investment, 182–83

tractor, mobile rabbit, 84–85

traits, selecting. *See* selection and genetics

treats, 65–66

tularemia, 40

U

utilities costs, 160. *See also* finances

V

veterinary costs, 160. *See also* finances

W

watering systems, 89–93

weaning kits, 56–57, 75

weather, protection from, 81–82

website marketing, 154–57

worm production, 13–14

PHOTO CREDITS

All photos are from the author except where noted below.

Photographers for the images from Shutterstock.com: KPiv and AMStudio_yk, front cover; Orest lyzhechka, pgs. 1, 17, 40; benchart, pgs. 3, 4, 8, 16, 42, 58, 76, 96, 112, 128, 140, 158, 178; Robyn Mackenzie, pgs. 7, 11, 19, 21, 26, 27, 31, 44, 46, 53, 56, , 81, 84, 92, 93, 111, 114, 115, 116, 118, 130, 132, 148, 150, 160, 170, 174, 176 (text background); Alberto Isidro Orozco, pg. 9; Everett Collection, pg. 14; Fototocam, pg. 13 (manure); Sukpaiboonwat., pg. 18; Julia Pivovarova pg. 20 (bottom); Kateryna_Moroz, pgs. 23, 100 (rabbit); Jarvna, pg. 19 (top); noypb, pg. 21 (bottom); vitrolphoto, pg. 34; Chanisa Ketbumrung, pgs. 32, 35 (top); Thanthima Lim, pg. 35 (bottom); Juli V, pg. 41; julia-kobzeva, pg. 30; Anca Popa, pg. 33; Sakan.p, pg. 29; Aubord Dulac, pg. 57 (top); OlhaSemeniv, pg. 52; PHATR, pg. 54; Roselynne, pg. 53 (top); Zozulia Mykola, pg. 44 (bottom); humphery, pg. 48; Anya Douglas, pg. 45; SGr, pg. 43; golfza.357, pgs. 65, 69 (feed); EricLiu1993, pg. 63; Tanakorn Akkarakulchai, pg. 66 (top); Daniele RUSSO, pg. 64; au_uhoo, pg. 66 (bottom); Ant Lees, pg. 59, 198; Blamb, pg. 62; Dmitry Chulov, pg. 67; WhiteYura, pg. 73; tomocz, pg. 90; mozrid, pg. 93 (bottom); WhiteYura, pg. 86 (top right); marilyn barbone, pg. 86 (top left); Stor24, pg. 78; noypb, pg. 89; Tinxi, pg. 104; Gorb Andrii, pgs. 109, 142; Eric Isselee, pg. 97; Vectorgoods studio, pg. 118 (bottom); AkaciaArt, pg. 113; Peredniankina, pg. 127; Ronald Rampsch, pg. 117; Anastasia_Panait, pg. 115 (top); unoL, pg. 135 (bottom); Todorean-Gabriel, pg. 131; createvil, pg. 129; Ajaykumar Mali, pg. 135 (top); cash1994, pg. 139 (USDA logo); Gorb Andrii, pg. 139 (rabbit meat); chinahbzyg, pg. 130 (bottom); Marian Weyo, pg. 150 (bottom); Elzbieta Sekowska, pg. 156; PPstock, pg. 155; nelea33, pg. 151; wow.subtropica, pg. 141; a katz, pg. 149; Dani Vincek, pg. 173; create jobs 51, pg. 159; DaCek, pg. 174 (top); Thongden Studio, pg. 177; Linn Currie, pg. 161; iceink, pg. 176 (top); vitec, pg. 172; JIANG HONGYAN, pg. 171; faithie, pg. 185; Jemastock, pg. 186 (farmers); ktasimar, pg. 183; Agnes Kantaruk, pg. 186 (roast rabbit); smereka, pg. 186 (buckwheat); Robyn Mackenzie, pg. 186 (rabbit cutlets); Xavier PT, pg. 182; Suto Norbert Zsolt, pg. 186 (honey); pelena, pg. 179; Leremy, pg. 186 (top left, top right); YeninMy, pg. 186 (sunflowers)

Ester van Praag, Arie van Praag, and Zahi Aizenberg, kindly provided the medical photos on dental diseases, sore hock, and ear mites.

ABOUT THE AUTHOR

Deborah Mays began her life with animals by grooming horses in exchange for free riding time. Volunteering in wild animal rescue secured her first zookeeper job at seventeen. She earned a degree in Wildlife Management, with elective courses that included animal physiology, pathology, parasitology, nutrition, and husbandry. Her husband of nearly four decades, John, loves to say that he found his wife in a cage in a zoo (which is actually the truth). Her career track was varied but always included animals, and she has worked with over 200 different species—from gorillas to tigers, alligators to giraffes, and rhinos to (of course) rabbits. She has worked in four different zoos, has been a veterinary technician and a wildlife biologist, and even rescued sea otters during the Exxon Valdez oil spill in Alaska. She spent 20 years as a Senior Research Specialist in biochemistry and cancer biology and authored 16 scientific papers. With John, she ran a commercial sheep and rabbit farm and an agricultural biotech company—all of which required intensive marketing and educational competence. Her book, *A Practical Guide to Rabbit Ranching*, is a merger of her passion for animals and science with her farm mantra: "You can't manage what you don't measure." She currently resides on Florida's Forgotten Coast, where she fishes, camps, shoots, kayaks, photographs wildlife, and writes.